Lightness, Brightness, and Transparency

Lightness, Brightness, and Transparency

Edited by
Alan L. Gilchrist
Rutgers University, Newark Campus

1994

LAWRENCE ERLBAUM ASSOCIATES, PUBLISHERS
Hillsdale, New Jersey   Hove, UK

Copyright © 1994, by Lawrence Erlbaum Associates, Inc.
All rights reserved. No part of the book may be reproduced in
any form, by photostat, microform, retrieval system, or any other
means, without the prior written permission of the publisher.

Lawrence Erlbaum Associates, Inc., Publishers
365 Broadway
Hillsdale, New Jersey 07642

Cover design by Kate Dusza

Figures 2.10 and 2.17 are reprinted from *Vision Research, 32*, P. Whittle, Brightness, discriminability, and the "Crispening Effect," pp. 1497–1497, 1992; Figure 2.16 from *Vision Research, 26*, P. Whittle, Increments and decrements: Luminance discrimination, p. 1687, 1986; both with kind permission from Elsevier Science, Ltd., The Boulevard, Langford Lane, Kidlington, OX5 1GB, UK.

Figures 4.3 and 4.4 are reprinted by permission of Academic Press.

Figure 6.3 is reprinted by permission from the *Journal of the Optical Society of America, 61*, pp. 1–11.

Figures 6.9, 6.10, 6.11, and 6.13 are from *Perception and Psychophysics, 28*, 527–538, reprinted by permission of Psychonomic Society, Inc.

Library of Congress Cataloging in Publication Data

Lightness, brightness, and transparency / edited by Alan L. Gilchrist.
    p. cm.
    Includes bibliographical references and index.
    ISBN 0-8058-0800-0
    1. Brightness perception. I. Gilchrist, Alan L.
BF241.7.L54   1994
152.14'3--dc20   94-3933
                                                                             CIP

Books published by Lawrence Erlbaum Associates are printed
on acid-free paper, and their bindings are chosen for strength
and durability.

Printed in the United States of America
10 9 8 7 6 5 4 3 2 1

To Jini

Contents

1 Introduction: Absolute Versus Relative Theories of Lightness Perception   1
Alan Gilchrist

   1. Introduction   1
      1.1. The Trieste Group   1
   2. The Relational Approach   2
      2.1. Features of the Relational Approach   3
      2.2. Spatial Frequency Analysis and the Photometer/Relational Distinction   13
      2.3. Does Relative Luminance Mean Contrast?   14
      2.4. Why Contrast Theories Are Not Relational Theories   15
   3. The Photometer Metaphor   17
      3.1. Symptoms of the Photometer Metaphor   18
   4. Mediating Processes from the Relational Perspective   24
      4.1. Edge Integration   24
      4.2. The Simplicity Principle   28
      4.3. Analysis of Components   29
      4.4. Classification of Edges   31

2 The Psychophysics of Contrast Brightness   35
Paul Whittle

   1. Contrast Brightness   35
      1.1. Introduction   35

1.2. A Little History	38
2. Matching Contrast Brightness	39
2.1. Outline of Experiments and Data	39
2.2. An Equation and an Ideal von Kries Scheme	44
2.3. The Brightness of Increments	45
2.4. The Darkness of Decrements	57
2.5. Matching Apparent Contrast	63
2.6. Flashing the Surrounds Instead of the Patches	64
3. Scaling Contrast Brightness: The Crispening Effect	64
4. Mapping Contrast Brightness Space	67
5. Discriminating Contrast Brightness	71
5.1. Measuring Discrimination Threshold	71
5.2. How Discrimination Threshold Varies with ΔL and L_b	73
6. Convergences and Divergences	77
6.1. How Much Convergence Is There Between the Three Methods?	77
6.2. Fechner's Integration	78
6.3. Brightness Functions	81
6.4. A Multiplicity of Contrast Signals?	84
6.5. Brightness as Just Another Contrast Dependent Response	85
6.6. What Produces Good Grays?	86
6.7. The Usefulness of $W = \Delta L / L_{min}$	87
6.8. Summary of Increment–Decrement Differences	88
6.9. Sources of Misunderstanding	89
7. Two More Types of Evidence	91
7.1. Dynamic Contrast Brightness	91
7.2. Contrast Color	92
8. Sensory Processes?	97
8.1. Contrast Brightness and Retinal Processes	97
8.2. Attenuation	98
8.3. Differencing	98
9. Explaining Contrast Brightness	101
9.1. Contrast Brightness Depends on Spatial, not Temporal, Differences	101
9.2. Is Discounting Explained by a Subtractive Component of Light Adaption?	102
9.3. A Role for the Brain?	105
9.4. A Two-Stage Model for Contrast Brightness	107
10. Conclusions on Contrast Brightness	108
10.1. An Ambiguity: Perception of Luminance or of Contrast?	108
10.2. Conclusions	109

3 Contrast Brightness and Ordinary Seeing   111
Paul Whittle

1. Introduction   111
   1.1. Contrast Brightness: A Special Case   111
   1.2. Plan of the Chapter   112
   1.3. Lightness and Brightness   113
   1.4. Ordinary Seeing?   114
2. Comparisons with Other Studies   115
   2.1. Candidates for Comparison   115
   2.2. Simultaneous Brightness Contrast   116
   2.3. Lightness Constancy   119
   2.4. Summary of the Comparisons   125
3. Two Patterns of Results and Two Types of Lightness Constancy   126
   3.1. Differentiation and Integration   126
   3.2. Equating Contrast   127
   3.3. Equating Luminance   128
   3.4. Experimental Evidence for the Two Kinds of Constancy   129
   3.5. The Perception of Lighting   131
   3.6. What Happens in Experiments on Brightness Contrast?   132
   3.7. Surround Reflectance Versus Illumination, and Lightness Versus Brightness   133
   3.8. Integration and Assimilation   134
   4.1. The Need for Integration: Recovering Luminance From Contrast   134
   4.2. Assimilation: A Manifestation of Integration?   136
   4.3. Is Seeing the Difference Between the Surround Luminances a Necessary Condition for Assimilation?   138
   4.4. Conclusions on Assimilation   140
5. Contrast Brightness: When and Why?   141
   5.1. What Conditions Produce Contrast Brightness?   141
   5.2. Why Contrast Brightness? Analysis Into Objects, Lighting, and Spatial Configuration   142
6. Contrast Brightness: Nature and Limitations   145
   6.1. The Phenomenology of Contrast Brightness   145
   6.2. The Paradox of Taking Background Brightness as Zero   147
   6.3. The Intransitivity of the Relation "Brighter Than"   147
   6.4. The Limitations of Contrast Brightness   150
   6.5. What Colors Are Seen in the HSD?   151
7. Explaining Simultaneous Brightness Contrast   152
8. Conclusions   154
   8.1. Summary   154

8.2. A Revised Partition, but not a Split, Into "Sensory" and "Perceptual" Color Perception? — 155

4 Surface Colors, Illumination, and Surface Geometry: Intrinsic-Image Models of Human Color Perception — 159
L. Arend

1. Introduction — 159
2. The Problem — 160
   2.1. Single-Illuminant and Multiple-Illuminant Scenes — 160
   2.2. Unasserted Color and Apparent Surface Color — 161
3. Failure of Unidimensional Constancy Models in Multiple-Illuminant Scenes — 162
   3.1. The Scene Segmentation Problem — 166
4. Intrinsic-Image Models — 166
   4.1. Computational Models — 167
   4.2. Inverse Optics — 169
   4.3. Human Perceptual Models — 172
5. Two-Stage Human Model — 173
   5.1. Stage 2, Inverse Optics Processes — 174
   5.2. Stage 1: Role of Sensory Mechanisms in Surface Perception — 181
   5.3. Psychophysics and Intrinsic Images — 194
6. Chromatic Surface Perception — 199
   6.1. Chromatic Variation of Illumination Within Scenes — 199
   6.2. Failure of Single-Dimensional Models: Color Appearance — 201
   6.3. Chromatic Intrinsic-Image Model — 203
   6.4. Sensory Processes — 211
7. Summary — 213

5 Achromatic Transparency — 215
Walter Gerbino

1. Introduction — 215
   1.1. Optical Ecology — 216
2. Layer Transparency and Color Scission — 217
   2.1. Layer Physical Model — 218
   2.2. Hierarchical Organization — 221
   2.3. Levels of Scission Theory — 226
3. The Episcotister Reflectance Model — 228
   3.1. Episcotister Constraints to the Scission Function — 230

3.2. Graphical Representation	232
3.3. Illumination Constraints	238
4. The Episcotister Luminance Model	240
5. The Filter Reflectance Model	245
6. The Filter Luminance Model	247
7. A Comparison of Episcotister and Filter Predictions	250
8. An Evaluation of the Scission Hypothesis	251

6 Color Constancy: Arguments for a Vector Model for the Perception of Illumination, Color, and Depth — 257
Sten Sture Bergström

1. Perceptual Constancy	258
2. Color Constancy and a Vector Model	259
2.1. The Main Assumptions of our Model	260
2.2. Retinal Illumination Analyzed	262
3. Empirical Support to the Commonality Assumption	265
3.1. One Field Versus Two or More Fields in the Same Illumination	265
3.2. Simulated Attached Shadows	266
3.3. One Rather Than Two Sources of Light	271
3.4. Simulated Illumination	272
3.5. Conclusion	272
4. Color Constancy and Depth Perception	272
4.1. Color Constancy and Perceived Solid Shape	273
4.2. Color Constancy and Perceived Separation in Depth	278
4.3. Conclusions About Color and Depth	280
5. Unecological Sources of Light	282
6. Our Model Summarized	283
6.1. Postulates and Assumptions	283
6.2. Edges and Gradients	284
6.3. The Minimum Principle	285
6.4. Why Call it a Vector Model?	285
7. Conclusions	286

Glossary	287
References	291
Author Index	309
Subject Index	315

1

Introduction: Absolute Versus Relative Theories of Lightness Perception

Alan Gilchrist
Rutgers University/Newark Campus

1. INTRODUCTION

In the fall of 1986 I asked each of the authors of this volume to come to the Fourth International Conference on Event Perception and Action in Trieste, Italy and participate in a symposium entitled "Absolute versus relative theories of lightness perception." I was not surprised that each of them agreed; the logic that connected our work seemed obvious to me, and I assumed it was equally obvious to the others as well. I discovered only much later that this was not true. It seems the others were puzzled at the composition of the group, one or two regarding it as crazy. In the meantime we all met in Trieste and had an unusually stimulating week together.

1.1. The Trieste Group

Before leaving I proposed to the others that we work together in certain ways, first by writing a book together, and then by trying to meet at least once a year for a few days to discuss questions of lightness perception. The others tentatively agreed and the following summer the five of us met for 3 days in a remote cottage in Wales. The next 2 years found us in a cabin in Lappland and then in a manoir in the Burgundy countryside. The discussions were intense and there was plenty of controversy. No one's thinking was able to remain unchanged. A common language began to develop, if not a common theory. Mind you, the locations were not hard to take; we were not in pain.

But by the Burgundy meeting we were so involved in our discussions, using these chapters as grist, that we completely neglected the sight-seeing we had planned, pausing periodically only for a succulent meal at one or another nearby village.

It was at this meeting that it became clear to me how little of my concept of the group was endorsed by the others, as they tore a draft of this chapter to shreds. Since then I have come to appreciate the differences among us, or at least the constructive role that those differences have played in fostering our work. The differences remain in these chapters; they have not been papered over. Yet there is a coherence that should not be missed. Here is my version of that coherence.

2. THE RELATIONAL APPROACH

I believe these five authors are among the best representatives of a change that has been taking place in the way achromatic surface perception is thought about. This trend might be called the relational approach. It is distinguished by more modern conceptions of both the input to visual processing and the output (i.e., what we see). These redefinitions of the input and output require, in turn, new concepts of the processes that intervene. In this view the input consists of relative luminance, not luminance per se, with an emphasis on edges and gradients. The output is multidimensional, in that every location in the image is perceived as having *at least* two perceived values, a surface lightness, and a certain degree of illumination. To account for these dimensions of visual experience, it is necessary to postulate processes that are not blind to the structure of the image, processes that relate disparate regions of the image by integrating luminance edges and luminance gradients, processes that classify these edges in terms of their physical cause. Ultimately, the approach is relative in the sense that perceived achromatic qualities are not locally determined but depend, at least potentially, on every other part of the visual field. This interdependence of different locations in the image can be seen in rudimentary form right at encoding. An edge or gradient is necessarily extended in space, whereas a point (which is in any case a hypothetical concept) is not.

Beyond this characterization, there is no set of theoretical concepts to which we all subscribe. The coherence of the group is seen most clearly in the comparison between our work and that of the more traditional approaches to lightness. I have used the term *photometer metaphor* to describe this traditional approach (Gilchrist, Delman, & Jacobsen, 1983). It refers to a perspective that has been common to almost the entire range of theories in the lightness/brightness domain, in spite of other important differences.

2.1. Features of the Relational Approach

I believe that the emerging relational perspective can be distinguished from the photometer perspective in terms of the following five comparisons:

1. Edge determination versus a point-by-point treatment of the image.
2. Relative luminance versus absolute luminance as the input.
3. Coding of information versus energy.
4. Multidimensional versus unidimensional view of the output.
5. A layered conception of the image versus a mosaic conception.

2.1.1. Edge Determination Versus a Point-by-Point Representation of the Image. The photometer perspective is closely associated with local determination of perceived qualities, the relational perspective with global or remote determination. The photometer metaphor can be thought of as a holdover of the old *constancy hypothesis*, according to which there is a constant relationship between local retinal stimulation and perceived quality. The almost forgotten term *constancy hypothesis* is confusing nowadays because its meaning is almost opposite to the more familiar use of the term *constancy* referring to constancy of perceived qualities despite changing proximal stimulation. Koffka (1935) noted that the constancy hypothesis:

> derived the characteristics of behavioural objects from the properties of local stimulation. In its consistent form the constancy hypothesis treats of sensations, each aroused by the local stimulation of one retinal point. Thus the constancy hypothesis maintains that the result of local stimulation is constant, provided that the physiological condition of the stimulated receptor is constant (e.g., adaptation). (p. 96)

Although models based on lateral inhibition are thought to explicitly reject the idea of local determination, I believe they ultimately do not. They depart from the constancy hypothesis in the sense that the rate of firing at a given point reflects more than just the excitation present at that point; it reflects the inhibition acting on that point from spatially removed points. But lightness is still defined on a point-by-point basis at a relatively peripheral level. This is shown in the incorrect prediction these models make (see Cornsweet, 1970, p. 350) that homogeneous surfaces will not appear homogeneous. According to this type of model, the spatial function of lateral inhibition requires that either bright or dark scallops will appear near the borders of homogeneous surfaces. This requirement is inconsistent with the simple fact that homogeneous surfaces generally appear homogeneous. But notice that it is not the inclusion of lateral inhibition per se in the model that requires the scallops; the problem lies in the point-by-point concept. Perception of the scallops is predicted only if one accepts the neural activation at each point (including the inhibition received at that point) as the correlate of perceived lightness, as Cornsweet does. In an alternative, more relational approach, the scallops do not appear in experience; they are left behind, as only the maximum/minimum differences at the border are forwarded to higher centers and the regions between borders filled in with homogeneous shades.

Whittle points out in chapter 3 that even from an edge-based approach there is the analogous problem of what homogeneous shade of gray a region is filled in with when the borders are not homogeneous in contrast, as with a typical element within a Mondrian pattern. This observation serves to remind us that the approach can be too local even when edge determination is assumed. It cannot be that surface color is simply the "filling in" of information from the border of the surface; relationships among many edges within the field must be considered. Discussion of this problem can be found in most of these chapters in terms of processes such as edge integration, edge classification, luminance relationships at intersections, and common and relative components.

2.1.2. *Relative Versus Absolute Luminance as Input.* Probably no one would argue any longer with the claim that perception of surface lightness is based on relative rather than absolute luminance. That was established most clearly when Wallach (1948) showed that a disk surrounded by an annulus appears equal in lightness to another such disk, not when the disks are equal in luminance, but when they have equal disk–annulus luminance ratios. Now the controversy has shifted to some closely related questions. Is absolute luminance information available at all to the visual system? Is relative luminance derived from absolute luminance? Does the visual system need to know absolute luminance?

Wallach himself assumed that luminance ratios were constructed out of absolute luminances coded by the retina. The discovery that images fixed in their retinal position disappear suggested the radical hypothesis that relative luminance is picked up directly by the movement of receptor cells across edges/gradients in the image, allowing a simpler interpretation of Wallach's findings. Attractive as this view is, it has been challenged by claims that high-contrast images (Kelly, 1979a, 1979b) and binocularly presented images (Rozhkova, Nikolaev, & Shchardrin, 1982) do not disappear when stabilized. However, Arend argued from his work with Timberlake (Arend & Timberlake, 1986) that the exquisite sensitivity of visual detection of temporal change combined with artifacts inherent in image stabilization techniques make it impossible to reject the possibility that a perfectly stabilized image, of whatever sharpness or contrast, would disappear completely.

The ganzfeld, or homogeneous visual field, allows a crucial test because it involves the presentation of light in the absence of edges or gradients. It has long been known that no surface lightness is perceived in a ganzfeld; in fact, no surface is seen at all. But the question has been raised as to whether the sheer intensity of light is sensed in the ganzfeld. Barlow and Verillo (1976) presented ganzfelds of different brightnesses and asked observers to estimate the brightness of each. They concluded that "the visual system is capable of performing as a photometer" (p. 1294). Unfortunately, their experiment leaves the obvious possibility that their observers were responding, not to the luminance of the ganzfeld per se, but to the luminance transient at the onset of each trial.

To avoid this problem Schubert and Gilchrist (1992) conducted an experiment in which the luminance of a ganzfeld was either increased or decreased at a constant very slow rate to see how soon observers could detect the direction of the change. Using a rate of 0.045 log units per min, or 9.5 times slower than the slowest rate of change previously found to be perceptible (Waygood, 1969; Yarbus, 1967), we found that observers detected the change in an average of 9.7 min, representing almost a threefold luminance change. In a second experiment we presented a disk within a ganzfeld. One of the two regions (either the disk or the ganzfeld) maintained a fixed luminance, whereas the luminance of the other was changed at the same slow rate as in our first experiment. In this case observers detected the direction of the change in an average of 1.02 min. In almost all cases the perceived change was attributed to the disk, regardless of which region was in fact changing. We concluded that a rough encoding of absolute luminance seems to exist. But relative luminance (on which surface lightness is based), because it is available so much sooner than absolute luminance, is probably encoded directly, not derived from absolute luminance information.

It has often been assumed that relative luminances are derived from absolutes, but this need not be the case. For example, the weight ratio between any two objects can be determined directly (i.e., without ever knowing the weight of either) by suspending them from opposite ends of a seesaw and bringing them into balance by adjusting the location of the fulcrum.[1] The weight ratio can then be determined from the ratio of the distances of each from the fulcrum. Likewise, luminance ratios of adjacent regions can be taken directly, without deriving them from pairs of absolute values. In fact, it is entirely possible that our ability to match the absolute luminance (technically, brightness) of two nonadjacent regions within an image is derived from an integration of all edges and gradients, both reflectance and illumination, that intervene between the two surfaces. Such an integral would correspond to the physical luminance ratio between the two surfaces, and perhaps this has misled us into thinking that the physical luminance is what is actually encoded.

The disks in Wallach's disk–annulus displays were matched for relative luminance, not absolute. But can they be matched for absolute luminance? The answer seems to be yes. Wallach obtained his ratio data only for decrements: disks darker than the surrounding annulus. When increments are used, one readily obtains luminance matching (Arend & Goldstein, 1987b; Jacobsen & Gilchrist, 1988b). But Arend and Goldstein (1987b) showed that one can also obtain ratio matching using increments, and crude luminance matching using decrements.

What does it mean that such center/surround displays can be matched either for absolute or relative luminance? From the photometer perspective it suggests that absolute luminances are encoded initially and then ratios are derived from these, allowing both ratio and luminance matching. From the relational perspective the argument might be that relative luminances are encoded initially, and then luminance matches are performed by integrating all the luminance ratios lying on a path connecting the two disks. So it turns out that matching experiments using center/surround stimuli can have two very different interpretations. How can we choose the more valid one? The argument that relative luminance is derived from absolute luminance is seriously weakened by our finding (Schubert & Gilchrist, 1992) that relative luminance is detected long before absolute. But an even stronger test was devised by Whittle and Challands (1969).

In Whittle and Challands' (1969) elegant experiments, observers matched center/surround stimuli using a unique haploscopic method. The two backgrounds were binocularly fused and, because the two targets were horizontally displaced away from each other, they appeared side by

---

[1]It might be argued that these two distances would have to be measured, and so the problem is merely shifted from measuring individual weights to measuring individual distances. However, one distance need only be measured in terms of its multiple of the other distance. Relative luminance can be treated in the same way. It seems necessary only that the visual system code an edge or gradient in terms of how many times brighter than the dark side is the brighter side.

side on what appeared, in the binocular view, to be a single background (see Fig. 2.1, p. 40). An integration of all the edges lying physically between the two targets was thus prevented because the edges of the backgrounds were not present in the binocular image. From the relational perspective this would have prevented luminance matching, allowing only ratio matching. From the photometer perspective, however, both kinds of match should still be possible because there is nothing to prevent the picking up of luminance information. In fact, throughout most of the photopic range, luminances cannot be matched at all under these conditions; only ratios can be matched. This outcome strongly supports the idea that the earliest encoding involves ratios.

Whittle's experiments, like those of Wallach (1948), Hess and Pretori (1894), and Heinemann (1955), probe brightness perception under minimal conditions. Yet Whittle's technique has the great advantage of distinguishing whether the ratio matching stems from a relative or an absolute input. This is why it is so significant that under these conditions there is a continental divide between increments and decrements. Brightness has long been treated as the phenomenal experience of the receptor input. But Whittle's work contradicts this idea, suggesting instead that sensed brightness is derived from an input that is highly relative. When one measures brightness under these conditions, subjects do not match increments with decrements.

With this technique so apparently well suited to measuring the earliest sensory quality, any increment appears brighter than any decrement, regardless of luminance. This is a remarkable fact, well worth emphasizing. These findings deal a strong blow to the conventional idea that relative luminance is derived from absolute luminance. I recommend a close look at Whittle's discussion (p. 146) of the question of whether brightness is a dimension with only positive values that increase from a zero at darkness, or a signed dimension with both positive and negative values increasing away from a zero of zero contrast.

The exclusivity of increments and decrements in early vision goes hand in hand with a relational perspective. It is precisely the relationship that changes at the boundary between increments and decrements. Increments and decrements have no special status under the assumption that luminance is encoded. Notice also that the concept of increment, just as that of decrement, involves a relationship between a photometric relationship (the luminance ratio, or at least its sign) and a geometric relationship (one region surrounded by, or adjacent to, another). Of course a relationship between two relationships is a higher order relationship. It is remarkable that such high-order relationships (increments and decrements) are so distinct in early vision.

If absolute luminance information is available at all, it seems to be fairly crude. What evidence do we have that perception of lightness and brightness requires absolute luminance information? As noted earlier, we do have some ability to compare noncontiguous regions of the visual field for brightness (perceived absolute luminance) apart from their perceived surface lightness. But this ability can be accounted for quite easily by edge integration (see p. 24). Perception of the overall (DC) level of illumination on the other hand does seem to require absolute luminance. Whereas relative luminance can account for perception of illumination variations within a scene, it would not appear to be able to account for the overall level. At the same time there are no data to indicate how accurately we perceive DC illumination level, and there is no reason to assume that this accuracy is any greater than Schubert and I obtained in our ganzfeld experiment.

Most of the data in this field are consistent with either the coding of relative luminance or the coding of absolute luminance. The most prominent results that have been claimed to require absolute luminance are contained in papers by Jameson and Hurvich (1961), Stevens (1961), and Hess and Pretori (1894). According to Jameson and Hurvich, these findings show the "intensity dependence" of lightness perception. Jacobsen and I reinvestigated these findings and concluded that they do not support the Jameson and Hurvich claim (Gilchrist & Jacobsen, 1988; Jacobsen & Gilchrist, 1988a, 1988b). The Hess and Pretori results change dramatically once a grievous methodological flaw is removed, the Jameson and Hurvich results simply cannot be replicated, and the graph of diverging functions presented by Stevens (1961), contrary to common assumption, represents a hypothetical experiment that was never in fact conducted (see Flock, 1974, p. 192; Jacobsen & Gilchrist, 1988a, p. 5).

2.1.3. *Emphasis on Information Rather Than Energy.*

The photometer metaphor implies that the visual system responds to the light itself rather than to relationships or changes in the light. Imagine that the letters on this page were printed in red ink rather than black. This should not prevent me from communicating to you the image of a green umbrella, because, as the reader, you derive the meaning from the pattern of the pigment, not from its color. If when you read the words "green umbrella" you form an image of a red umbrella simply because the words "green umbrella" are printed in red ink, you would be (besides showing a reverse Stroop effect) confusing the energy with the information. Focusing on relative luminance treats light as the vehicle and variations in its intensity as the information carried.

2.1.4. *Multidimensional Output, not Unidimensional.*

When we are speaking of the output of the visual system, we are speaking of phenomenology. We need to take this seriously. Because a prime aim of sensory physiology has been to place visual perception on a firm materialistic foundation (the embarrassing failure of introspectionism has not been forgotten), phenomenological considerations have sometimes been given short shrift. The problem, however, is that visual experience is precisely what a theory of perception is charged to explain. The facts of phenomenal experience, even though measured behaviorally, are the only facts by which our theories can be tested. Thus phenomenology cannot be avoided; it is of the first moment that visual experience be fully and correctly described.

Visual experience tells us that perception is multidimensional, a fact that presents an enormous challenge for any point-by-point approach. Even allowing the value at each point to be influenced (e.g., through lateral inhibition) by luminance values at nearby locations does not help much because one is still left with a single value at each location. This leaves a one-to-many mapping problem because that single point will have two or more perceived values. For instance, besides having a surface lightness, it will be illuminated more or less brightly, and it may even be seen to lie behind a veiling luminance of a certain brightness or a transparent layer that may have its own lightness and illumination. In a relational model, conversely, all the points falling within a closed edge or gradient are defined by the nature of that edge (although more remote edges also play a role). Because a given point typically falls within more than one closed contour, it can easily have more than one value.

Consider a concrete example: a gray square on a white background with a shadow falling on half of the gray square and its background. All the points within the gray square share a common value of surface lightness that is determined by the luminance change at the border of the gray square, and the relationship of that border to other reflectance borders in the larger context. All the points lying within the shadowed half of the gray square share an additional perceptual property: They are perceived to be dimly illuminated. This property, shared as well by all the points lying within the shadowed part of the white background, is determined by the luminance change at the border of the shadow, and the relationship of that border to other illumination borders in the larger context. The points within the shadowed half of the gray square thus have two perceived properties, just as they are surrounded by two edges. Each property, the gray surface color and the dim illumination, is associated with one of the two surrounding edges. Likewise, in a Venn diagram, a point falling in the region of overlap between two circles labeled *doctors* and *New Yorkers* will have two separate values, corresponding to the definition of the two circles.

Indeed, constancy often fails just because reflectance and illumination edges are inseparable. The Gelb effect is a well-known example. A disk of black paper is suspended in midair and illuminated by a special beam of light that illuminates only the disk. The disk appears white because the border of the black surface color is coincidental in the image with the border of the illumination. One might say the two edges are confounded. There are of course cases in which constancy survives the coincidence of a reflectance edge and an illumination. This may occur at a corner where walls of two colors, differently illuminated, meet. But in such cases the larger context allows the relative contributions of lighting and surface color at the corner to be disentangled.

The dimensions of visual experience we have been considering represent multiple objective dimensions in the physical world that is seen. All these perceived dimensions fall within the normal object-oriented way of seeing that Rock (1975) called the constancy mode. But visual perception is multidimensional in another sense. There is a mode of perception that is used by painters but is more or less accessible to all. Rock called it the proximal mode; Katz (1935) called it the subjective attitude; MacLeod (1932) called it photographic vision; Gelb (1929) called it an "attitude of pure optics." It is an attitude that, according to Köhler (in Katz, 1935), causes us "to free ourselves from the object-character of the surfaces and degrade them into mere extents of light" (p. 167). In this attitude, we see the combined effect of surface lightness and illumination. The term *brightness* has come to refer to the seeing of this combined effect. One might say we see the energy rather than the information, but it is not a normal way of seeing (except for seeing light sources), and our perception of brightness is not as well correlated with light levels as our perception of lightness is correlated with surface gray colors. Katz' classic work, *The World of Colour* (1935), unsurpassed in its treatment of the multiple dimensions of visual color experience, gives an extensive analysis of these various dimensions.[2]

The lightness/brightness distinction has been applied in at least one important way that is different from the preceding definition. For example, Koffka (1935) and Rock (1975) used the term *brightness* to refer to the perceived level of illumination. Today this usage is incorrect, not because it does not refer to a valid perceptual distinction but simply because it is not consistent with the consensus regarding usage that has emerged.

---

[2]Because there has been some confusion with regard to some of the dimensions Katz described, it may be useful at this point to place two of his terms, *insistence* and *pronouncedness*, in the context of the previous discussion. Insistence is essentially equivalent to *brightness*, and, for a given surface lightness, it correlates positively with illumination level. Pronouncedness, on the other hand, does not vary so simply with illumination. According to Katz, whites are more pronounced in higher illumination, whereas blacks are more pronounced in lower illumination (see especially p. 81).

Using computer simulation Arend presented a pair of Mondrian patterns that differed only in their illumination level (Arend & Goldstein, 1987b). Observers were required to match a target in one Mondrian with a target in the other Mondrian, either for lightness or for brightness. He obtained very good reflectance matching in the lightness task and brightness matches that varied with luminance. The importance of his results is that they show that observers can make two very different kinds of match for the same stimulus. Jacobsen and I (Jacobsen & Gilchrist, 1988a) showed the same thing for a simpler Mondrian presented successively under three levels of illumination. Arend (1990) subsequently showed, using center/surround patterns embedded within Mondrians, that observers can match targets on the basis of their local contrast ratio, independent of both lightness and brightness matches. My students and I showed (Gilchrist, Delman, & Jacobsen, 1983; Gilchrist & Jacobsen, 1984) that observers can match the level of illumination on surfaces separately from their matches of lightness.

Gerbino and his colleagues (Gerbino, Stultiens, Troost, & de Weert, 1990) demonstrated transparent layer constancy using a technique similar to that used by Katz to demonstrate lightness constancy. Observers were presented with a pair of patterns side by side, each consisting of the simulation of a transparent layer in front of a simple checkerboard pattern. They were asked to match the appearance of the two transparent layers by adjusting certain luminances in the comparison display. Observers were quite successful at this task. Gerbino showed the interplay between surface lightness, transparency, and perceived illumination, arguing that each surface we see is experienced as having a value on the transparent–opaque dimension as well as separate values on the more familiar black–white or glossy–matte dimensions.

My work with Jacobsen on lightness constancy through a veiling luminance (Gilchrist & Jacobsen, 1983) illustrates a hitherto-neglected case of multiple perceived dimensions. Observers show remarkable lightness constancy when viewing three-dimensional scenes through veiling luminances that dramatically reduce luminance contrasts, from 30:1 down to 2:1. It is interesting that lightness constancy in these experiments always goes hand in hand with perception of the veil. Under conditions where the veil is not perceived, constancy is not obtained. Again we find multiple perceived values at each point in the visual field. I once conducted an experiment in which one image reflected in a sheet of clear glass was superimposed on another image seen through the glass. The images were arranged so that a red object in one image overlapped a green object in the other image. Intensities were adjusted so that the red light and the green light should have produced a yellow appearance, according to the laws of additive color mixture. Yet not one of the nine observers reported perception of any yellow, and none could find any yellow color in the display when asked. A second group of nine observers viewed the region of red–green overlap through a hole in a screen (called a reduction screen) that occluded the remainder of the display. All nine of these observers reported the color in the hole to be yellow, and none could perceive any red or green. Just as the Gelb illusion is produced by a coincidence of the reflectance edge and the illumination edge, so in this case the hole destroys the separate red and green experiences by making the boundary of the red region coincide with the boundary of the green region. Apparently, the mixing of red and green light, as light, is not sufficient to invoke the law of color mixture. Except for very simple cases, the borders must be mixed as well.

Whittle has typically worked with displays so reduced that they allow only a single dimension of perceptual experience. But in his writing, particularly chapter 3, he discusses the various dimensions of perceptual experience (see especially p. 114). The ingenious haploscopic paradigm he developed locks the observer into a single dimension of experience. Yet, because this is achieved by preventing the visibility of the separate background borders, it suggests that other dimensions of experience that occur in more representative scenes depend on the integration of chains of edges. So his paradigm has, ironically, contributed to our understanding of multiple dimensions of experience by its success in excluding them.

2.1.5. A Layered Versus Mosaic Conception of the Image. The traditional view of the retinal image implies an image composed of adjacent points, as in a pointillist painting, or regions, as in a mosaic. The relational approach treats the image as composed of overlapping layers (see Fig. 1.2). In general, we do seem to perceive a pattern of illumination superimposed on a pattern of surface colors, as in the earlier example of the shadow falling across the gray square on the white background. Sometimes the layers are even more distinct perceptually, in perceived transparency situations or in my example of the superimposed images containing red and green objects. Neisser and Becklen (1975) once demonstrated that observers could selectively attend to either of two action films superimposed on one another, and the unattended one dropped out of consciousness altogether.

The idea of overlapping layers of light is there in work at the sensory end of perception by Whittle and by Walraven (1976). Bergström's decomposition of the image into components of reflectance, illumination, and three-dimensional form fits in well with the concept of layers, as does Barrow and Tenenbaum's (1978) concept of intrinsic images. Arend, Gerbino, and I have all talked in one way or another about superimposed perceptual layers. The concept of a layered retinal image fits comfortably with the multiple dimensions of visual experience.

2.2. Spatial Frequency Analysis and the Photometer/Relational Distinction

It is ironic that after 25 years of Fourier/spatial frequency analysis (based on the notion that vision is seeing contrast) there should be a need for a book stressing the importance of relative luminance as an input. When I speak of the photometer metaphor, I am specifically referring to the tradition within psychology that has grappled with the perception of surface blacks and white. The Fourier approach, although it has something to say about brightness perception, has not really attempted a theory of surface lightness. Despite this lack of overlap, some discussion of the relationship of the Fourier/spatial frequency approach to the photometer/relational distinction seems to be in order. Rejecting as it does a point-by-point analysis of the image, the Fourier approach cannot be identified with the photometer approach, and the Fourier emphasis on contrast provides a clear recognition of the importance of relative luminance. As for the coding of energy versus information, it is difficult to categorize the Fourier approach; both energy and signal metaphors are used.

Likewise it is difficult to identify the Fourier approach with either a unidimensional or multidimensional view of the output because there is so little treatment in the spatial frequency literature of the phenomenal experience of objective properties of the visual world. To be sure, the different spatial grains in the Fourier analysis could be considered as different dimensions. But it is not easy to see how one gets from these dimensions to the dimensions of visual experience. I am reminded of Wertheimer's (1923) critique of the introspectionists' hypothesized points of light: "I stand at the window and I see a house, trees, sky. And I could, then, on theoretical grounds, try to sum up: there are 327 brightnesses (and tones of color). (Have I 327? No: sky, house, trees; and no one can realize the having of the 327 as such.)" I believe Wertheimer would as easily have said: "Have I this amount of energy at 2 cycles per degree and that amount of energy at 6 cycles per degree? No, I have sky, house, trees."

What of the analysis of the retinal image into mosaic elements versus layers? The analysis of the retinal image into spatial frequency components is distinctly different from a mosaic conception of the image (as illustrated, for example, in a pointillist painting). But it also seems different from the analysis of the image into layers. Of the kind of layers postulated by Bergström, Arend, Gerbino, Gilchrist, or Barrow and Tenenbaum (1978), each can itself be analyzed into a set of spatial frequencies. Although the analysis into spatial frequency components allows the precision of powerful mathematical tools, the components do not map well onto the components of visual experience. We perceive a pattern of illumination superimposed onto a pattern of surface colors. There has been little success in parsing these two patterns using spatial frequencies. Even the attempt to associate reflectance borders with high spatial frequency and illumination borders with low spatial frequency receives a sharp critique in Arend's chapter.

2.3. Does Relative Luminance Mean Contrast?

The relational perspective can be regarded as either consistent with contrast or antithetical to it, depending on the usage of the term. So tortured has the term *contrast* been that we must first take a moment to sort out its various meanings. It will be sufficient to distinguish three predominant uses of the term (see also the glossary): as a measure of the stimulus, as an illusory phenomenon, and as an explanatory tool.

1. The term *contrast* can be used to refer to the relative luminance at an edge or gradient. Two frequently used definitions are Weber contrast ($\Delta L/L$), or Michelson contrast ($Lmax - Lmin / Lmax + Lmin$). This use, which is theoretically neutral, plays a crucial role in the relational perspective.

2. Contrast often refers to a particular perceptual phenomenon, the illusory effect on a surface of its background luminance. The most common example, often illustrated in textbooks, is called simultaneous lightness contrast and is typically composed of two identical pieces of gray paper on adjacent white and black backgrounds. The gray on the white background is perceived as slightly darker than the other gray. In this usage, contrast is simply the name for a phenomenon and is theoretically neutral.

There are also illusory effects that occur under more reduced conditions, that is, with a simple center and surround in a dark environment. One effect, also sometimes called simultaneous contrast, is that an increase in the luminance of the surround is perceived as a decrease in the brightness (or lightness) of the center, within certain parameters. This transfer of effect from surround to center, like induced motion, is an undisputed fact. It is sometimes called brightness induction.

Second, it is claimed that under these same conditions, the perceived difference between the center and the surround is increased (Heinemann, 1972, p. 176), magnified (Hurvich & Jameson, 1966, p. 85), or amplified (Cornsweet, 1970, p. 300). This supposed process of "edge enhancement," although widely assumed, has never been established. As Freeman (1967) wrote in a review of the contrast idea: "an experimental analysis of enhanced brightness differences has not, as yet, been performed" (p. 167).

Two distinctions must be clearly made, however. First, the "edge enhancement" idea and the "transfer of effect," or induction, idea, are independent. One does not imply the other. Increasing the luminance of the surround can cause a perceptual darkening of the center even as the perceived relationship between the center and surround remains fully consistent with the physical luminance ratio.

Second, the simple center/surround display, which I call the laboratory display, should not be confused with the textbook display. Although induction may occur in both displays, its magnitude is much greater in the laboratory display. I believe this is because, under such reduced conditions, the surround plays the role of the reference white, just as a surrounding framework plays the role of the stationary anchor in motion displays (Gilchrist & Bonato, 1993). This is revealed by the fact that the surround continues to appear white, regardless of its luminance. When the luminance of this reference white increases, a center region of constant luminance will appear darker in direct proportion. In the textbook example there is no shift of the white level; the two gray targets are related essentially to the same reference white, generally the white paper of the book.

3. Contrast frequently refers to a hypothesized neural process, usually identified with lateral inhibition, by which the excitation produced in a focal region is reduced by neural activity in an adjacent or surrounding region. Various theories (Cornsweet, 1970; Jameson & Hurvich, 1964) have proposed to use such a neural process to account for phenomena such as lightness constancy and simultaneous lightness contrast. I refer to these theories as contrast theories.

I want to draw a particularly sharp distinction between the uses in items (1) and (3) here. All the authors in this book stress the importance of relative luminance, either luminance differences or luminance ratios. We sometimes use the familiar expression *luminance contrast*. But what is implied is only the benign idea that some kind of relative luminance is coded at input. This does not necessarily imply any process of exaggeration of differences, or inhibition or induction. Those concepts are associated with what I call contrast theories, as in item (3) previously mentioned, and I try to make this clear by using the term *contrast theory* when I refer to such induction processes.

2.4. Why Contrast Theories Are Not Relational Theories

Contrast theories do emphasize the relationship between two luminance values, that of the center and that of the surround. But they seem to me importantly different from what I am describing as relational theories, for these reasons among others:

1. In contrast theories the correlate of perceived lightness is an amount, not a relationship. To be sure, that amount can be influenced by other

amounts, such as background luminance, but it still comes down to an amount. For example, Cornsweet (1970) defined the correlate of perceived lightness as "the frequency of firing of the spatially corresponding part of the visual system (after inhibition)" (p. 303). As Koffka (1935) wrote, "contrast . . . implies an explanation not in terms of gradient, but in terms of absolute amounts of light" (p. 245).

2. Also there is the idea of distortion. The amount in the target region is held to be irreversibly changed or distorted by the background luminance, or the difference in luminance at a border is held to be exaggerated.

I may be short relative to a giraffe, yet tall next to a chicken. But that does not mean that when a giraffe stands next to me my body actually shrinks. In a contrast theory, the presence of the background changes the value of the target surface itself, whereas in a relational theory, the presence of the background merely changes the value of the target/background relationship. Contrast theories speak of information loss (e.g., Cornsweet, 1970, p. 379), whereas relational theories speak of recoding of the input.

Different perceived qualities can depend on different relationships. A target luminance may indeed form a relationship with the background luminance, but at the same time, for other perceptual purposes, it may form one or more separate relationships with other parts of the perceptual field. For example, in a lightness constancy display using side-by-side illuminated and shadowed fields, it has been shown (Arend & Goldstein, 1987; Henneman, 1935; Jacobsen & Gilchrist, 1988a) that observers can match illuminated and shaded targets both for lightness and for brightness (perceived luminance). Further, there is good evidence that both lightness and brightness are based on relative luminance, albeit different combinations of relative luminance. Lightness seems to be based on an integration (see p. 31) of all the reflectance edges between the two targets, excluding any illumination edges (Arend & Goldstein, 1990; Gilchrist, 1988; Gilchrist, et al., 1983). Brightness seems to be based on a simple integration of all the edges between the two targets, reflectance and illumination. If brightness is based on luminance ratios rather than simply on luminance itself, then the same luminance ratio will often be used in separate processes (i.e., lightness and brightness). If the difference at the border of a target is distorted (by a contrast process) in the service of the lightness match, this alters one of the component relationships needed for the brightness match.

3. Contrast theories seem satisfying because they appear to account for the role of context. For example, a target surface of constant luminance will appear to get darker when the luminance of the surrounding region increases, regardless of whether the increase is perceived as due to a change in the reflectance of the background, a change in the illumination

(on both target and surround), a change in the intensity of a veiling luminance in front of the target, or a change in either the lightness or the transparency of a transparent layer in front of the target. But the magnitude of this effect—the amount by which the lightness of the target decreases—is different in each of these cases. And the difference can be dramatic. If the reflectance of the background is perceived to increase, this will have only a small effect, but if the illumination level of the background is perceived to increase, the effect will be much greater. In fact, if the reflectance of the background increases by a factor of 30 (i.e., from black to white), a surrounded target will appear only about one Munsell step darker, but if the illumination on the background increases by a factor of 30, the target will appear to darken by about six Munsell steps (Gilchrist, 1988). Remember that in both these cases, the same thing is happening to the retinal image in the region of the target and its background. The contrast idea is a minimal acknowledgment of the relational nature of lightness. Contrast theories are "structure-blind" and relational theories are not (or need not be). For instance, a retinal contrast mechanism has no way of distinguishing whether the luminance gradient that stimulates it comes from a change in surface reflectance in the environment or a change in the illumination level on a surface.

My own view is that the contrast mechanism has played a vital role in maintaining the photometer metaphor, by appearing to deal with the variety of context effects on lightness. Without contrast as an adjunct tool, the photometer metaphor would stand exposed as wholly inadequate. Contrast gives it plausibility. Thus Jameson and Hurvich (1961) could speak of the "intensity dependence" of the lightness system, although claiming to have a relativistic theory.

3. THE PHOTOMETER METAPHOR

Of course, no one explicitly endorses the photometer metaphor. The concept has been so widely shared as to be taken for granted. I believe we have been too close to it to see it. As Koffka (1935) observed of the constancy hypothesis: "I doubt whether at the present moment one would find a psychologist who would defend it explicitly, but that is not equivalent to saying that it has disappeared. The contrary is true. All applications of the interpretation theory contain it in some form or other" (p. 87). Likewise I argue that the photometer metaphor is revealed by the way in which it makes sense out of both the history of lightness theories, and of some observations about this field that would be merely puzzling in the absence of such a dominant paradigm.

3.1. Symptoms of the Photometer Metaphor

3.1.1. Similarities and Differences Among Theories. I argue that until recently theories of lightness have been so unanimous in their assumption of the photometer metaphor that they have been most clearly distinguished by the various corrective mechanisms that are proposed by each to transform the absolute luminance input values into perceived lightness values. Helmholtz (1866/1962), for instance, insisted that the photometer readings required an illumination frame of reference before lightness could be determined. Hering (1874/1964), besides noting that adaptation and pupil size reduced the constancy problem, emphasized the role of lateral inhibition in altering the photometer readings. Helson (1943) operationalized Helmholtz's frame of reference in his adaptation level. These were all models for processing luminances in one way or another. Cornsweet (1970) and Jameson and Hurvich (1964), echoing Hering, added an inhibition process to the photocell mechanism (called the excitation process by Jameson and Hurvich), differing mainly on whether the two processes either exactly (Cornsweet) or approximately (Jameson & Hurvich) cancel each other when illumination changes. Beck (1972), perhaps the most eclectic of all, added both lateral inhibition and cognitive judgments to the basic photometer readings. Even Wallach, who introduced the ratio idea in his landmark experiments, conceived the ratio as a ratio between two absolute luminances, independently given. Of course, such a ratio concept does not require absolute luminance information at all, as was noted earlier in the seesaw example.

Beck (1972), in his book on surface-color perception, expressed the familiar view that sensory and cognitive theories represent the theoretical poles of explanations of lightness perception. The prototypical sensory theory is probably the opponent process theory of Hurvich and Jameson (1966) based on the concepts of Hering (1874/1964). Yet they, like Hering, claim only to have a model of sensed brightness, not one of surface lightness, making it clear (pp. 102–104) that the disentangling of surface reflectance from illumination is a matter of cognitive interpretation, not sensation. This makes their theory strikingly similar to that of the prototypical cognitivist, Helmholtz. The difference is that Hurvich and Jameson put both an excitation process and an inhibition process into the sensory mechanism, whereas for Helmholtz there is only an excitation process. But both theories require a cognitive judgment before a specific shade of black or white can be selected, and for both theories that cognitive judgment is based on a sensory core. Similarly, MacLeod (1932) observed that Hering's "position is only slightly different from that of Helmholtz; he has simply attributed more to peripheral and less to central factors. The constancy of colour is due primarily to conditions in the receptors; but central modification may take place, e.g., in the case of memory colour" (p. 22).

In my view the more important theoretical watershed is between gestalt theory and those, whether sensory or cognitive, that involve the interpretation of sensations. Although gestalt theory is sometimes placed in the cognitive category, Köhler (1915) conducted his constancy experiments (including lightness constancy) with lower animal forms specifically to demonstrate that constancy does not require cognitive judgment. Koffka (1935) observed regarding size constancy that: "Chicks must be geniuses if they can discover in the first three months of their lives that something that looks smaller is really bigger" (p. 89). Gelb (1929) put it plainly:

> Our visual world is not constructed by "accessory" higher (central, psychological) processes from a stimulus-conditioned raw material of "primary sensations" and sensation-complexes; rather from the very beginning, the functioning of our sensory apparatus depends upon conditions in such a way, that according to the external stimulus constellation and the internal attitude we find ourselves confronted by a world of "things", thus or thus, now more poorly, now more richly articulated and organized. With this articulation and organization such aspects of the visual world as "visibility of a certain illumination", "existence of flat colour surfaces" and of "rich articulation of the field of vision" are intrinsically connected, and these aspects are the necessary conditions for the arousal of the constancy phenomena." (p. 673)

Koffka (1935), who of the three founders wrote the most on the topic of lightness perception, bypassed the excitation concept entirely. He explicitly considered lightness perception to be based from the beginning on gradients of light.

3.1.2. Neglect of Illumination Perception.

There has been very little work done on how we are able to perceive the level of illumination on various surfaces. Most theories either ignore the question entirely or deny illumination perception outright. Hering (1874/1964) said: "Most people notice even large differences in two illuminations only when they are presented side by side or in rapid succession" (p. 14). Cornsweet (1970) wrote that "our perceptions are correlated with a property of objects themselves (i.e., their reflectances) rather than with the incident illumination" (p. 380). This is very strange. Common sense tells us that we do have reasonably good perception of illumination. Shadowed areas appear very different from illuminated areas within a scene. And even before we sit down to read, we are very aware of whether or not there is enough illumination. And there is experimental evidence that illumination perception is actually quite good (Gilchrist et al., 1983; Gilchrist & Jacobsen, 1984). And of course shape-from-shading is based on gradients of effective illumination, not reflectance gradients. It could not work if the gradients were not well perceived.

Cornsweet (1970), for example, gave an analysis of how the rate of firing of cells exposed to a patterned array could remain invariant after an overall change of illumination on the array. But as Rock once pointed out to me, this model explains too much! It is a model of lightness identity, not lightness constancy. We are not talking about a metameric match. A gray surface in bright illumination and a gray surface in shadow do not appear identical in all respects, only with respect to surface reflectance. One gray still appears brighter than the other. This illustrates some of the confusion that can result from insufficient attention to phenomenal experience. The photometer metaphor has encouraged neglect of illumination perception. According to the photometer metaphor, each point in the visual field is regarded as having only one value. That value will tend to be given to surface lightness, leaving nothing for the perception of illumination.

There is a bizarre version of this story that enjoys more support than its faulty logic would suggest. For example, Helson (1943) wrote: "The failure in this case to achieve complete constancy serves as a cue to the fact that the illuminations are different" (p. 249). And, according to Jameson and Hurvich (1989), if lightness constancy were: "a perfectly compensatory mechanism, then there would certainly be no need for experience with, or judgments of, different levels of illumination, because their effects, at least for uniform illuminations and diffuse object surfaces, would never be registered at all beyond the most peripheral level of the visual system" (p. 5). Hering (1874/1964) wrote: "Without this approximate constancy . . . If the colors of objects were to be continuously changed in this way along with the illumination changes, then . . . *the momentary color in which a given object appears to us could serve as a criterion for the intensity or quality of its illumination*" (p. 17). Friden (1973) argued that the more conditions favor lightness constancy the more they produce insensitivity to illumination.

The argument common to these three quotations is that changes in the illumination level are revealed by failures in the constancy of surface lightness. This is nonsense. When lightness constancy fails, we do not experience it as a failure. In fact, for all we know, we may be the victims of widespread failures of constancy. We would need an independent measure of the physical reflectance to determine the degree of failure of lightness constancy. Hence there is nothing in our experience of surface lightness to signal a change of illumination. Even when a change of illumination seems (by mistake) to be accompanied by a change of surface lightness, we merely experience it as a change of lightness.

The implication is that the perception of illumination and the perception of surface lightness are locked in a zero-sum game. If lightness constancy were perfect, perception of illumination would be zero; if lightness constancy were zero, illumination perception would be perfect. In fact, the correlation between lightness constancy and perception of illumination is positive, not negative, as many studies have shown. As Bergström argues in his chapter, seeing the illumination is a condition for proper surface color perception, not an alternative to it. The logical failure here involves an attempt to force surface lightness and perceived illumination into a single dimension, a hallmark of the photometer metaphor.

3.1.3. Neglect of Stabilized Retinal Images.
The fact that stabilized images disappear is a remarkable fact, yet one that is not obviously consistent with the photometer metaphor. Holding the image steady should, if anything, make it easier to read the luminance at each point. I believe the relative neglect of this fascinating topic, not to mention its denial, can be understood in terms of the strength of the metaphor.

3.1.4. Status of Jameson and Hurvich's Diverging Functions Claim.
A 1961 study claiming that the strength of the ratio effect in turn depends on absolute luminance levels has been reported in almost every perception textbook, despite the fact that it has never been successfully replicated and there are four published failures to replicate (Flock & Noguchi, 1970; Haimson, 1974; Jacobsen & Gilchrist, 1988a; Noguchi & Masuda, 1971). Nor does this study stand up well to scrutiny. The display was produced by rear-illumination that makes it very difficult to maintain constant ratios (a key condition) because any stray light produces a veiling luminance on the screen, reducing contrasts in the projected image (Gilchrist & Jacobsen, 1983). There was very little description of the apparatus and method, and no statistical tests were presented. Yet this study was very effective in undermining the then-recently published Wallach ratio results.

The diverging functions concept undermines the ratio perspective. Given with one hand yet taken with the other, the ratio principle becomes no longer a general principle but merely an interesting curiosity found at certain luminance levels. Most of the data in this field are consistent with either relative or absolute luminance as an input. The central point of the Jameson and Hurvich experiment is that their data demand a visual system responding to absolute luminances; hence their term *intensity-dependence*. Apart from the crucial support this study provides for the dominant metaphor, it is hard to see why it should have acquired the status it has.

3.1.5. Neglect of "Background-type" Constancy.
Whittle (Whittle & Challands, 1969) and others (Brenner, Cornelissen, & Nuboer, 1991; Maloney & Wandell, 1986; Mausfeld & Niederee, 1993) pointed out that, besides lightness constancy with changing illumination (call it Type I), there is another crucial type: lightness constancy with changes in the background reflectance (Type II). To be sure, this type of constancy is not 100% complete. No constancies are. But it is every bit as strong as the more familiar constancy with respect to changes of illumination. I compared these two

types of constancy under comparable conditions (Gilchrist, 1988), by presenting targets on side-by-side backgrounds that appeared to differ either in surface color (black vs. white) or level of illumination. Observers matched the two targets for lightness, with the result that the matches fell equally short of what would constitute perfect constancy in the two conditions.

This newly discovered constancy is just as valid ecologically as the more familiar kind. We regularly see a surface against changing backgrounds. As Whittle (this volume) put it so strikingly:

> An object moved over different backgrounds does not seem to change much in lightness. To get a good look at a sample of cloth, you may pick it up and take it to a good light, but you do not worry about what background is behind it. It is as though the background does not matter. It is not that there is no simultaneous contrast effect, if you look for it; just that it is amazingly small. (p. 128).

He suggested that, without this second type of lightness constancy, "such objects would flash on and off all the time as they changed from being increments to decrements. They do not." Why then has constancy with respect to changing illumination been known and studied for more than a hundred years, whereas constancy with respect to changing background has only lately been noticed?

I believe this oversight flows directly from the photometer metaphor. If the sensory mechanism codes luminance, then constancy of surface lightness under changing illumination (Type I) becomes a problem because the luminance of a surface changes. Constancy of surface lightness with changing backgrounds (Type II) does not appear problematic because the luminance of the surface does not change. But if the sensory mechanism codes luminance ratios, then Type II becomes the obvious problem (because the luminance ratio changes) and Type I recedes. In fact, had the coding of luminance ratios been the accepted view from the beginning, there is no reason to think that constancy of surface lightness under changing illumination would have ever been seen as a problem.

Because of the photometer metaphor, even the introduction of contrast theories did not clarify the situation. According to contrast models, the lightness of a surface is supposed to change when its background changes, and lo and behold, it does (however modestly). Thus the importance of Type II constancy is even further obscured.

Both Whittle and Arend (this volume) discuss this kind of constancy in their chapters.

3.1.6. Form of the Constancy Question.

Not only has the photometer metaphor focused thinking on only Type I constancy, but it has distorted the analysis of even that one. MacLeod (1940) suggested that the very term *constancy* implies something like what I have called the photometer metaphor:

> There are several reasons, however, for considering the concept of constancy as a superfluous, if not a dangerous, concept in any scientific treatment of the problems of perception. In the first place, the principle as formulated implies the very hypothesis which it has helped to demolish, namely that of a primary correlation between stimulus and sensation. Having relinquished this hypothesis, it is misleading for us to continue to talk in terms of discrepancies between the physical and the phenomenal (p. 17).

In its traditional form the constancy question asks: "How can perceived lightness remain constant when the luminance of the surface changes (due to a change of illumination)?" Because this form of the question is inadequate, it has led to inadequate answers. For example, Wallach's ratio principle represents a fundamental insight that appeared to solve Hering's paradox, that is, how can two perceived variables, reflectance and illumination, be extracted from a single proximal variable, luminance? According to Wallach, surface reflectance (lightness) is given directly by the ratio rule. Once both luminance and reflectance are known, illumination can easily be derived. Yet this account works only when applied to reflectance edges, not illumination edges. With no mechanism for distinguishing these edge types, it fails to solve the paradox in the larger sense, that is, how are changes in illumination distinguished from changes in reflectance?

Contrast theories answer the traditional question by asserting that, although the luminance of a target in bright illumination may be greater than that of an equally gray target in dim illumination, lateral inhibition equalizes their neural excitation. The higher neural activity associated with the target of higher luminance is presumably more strongly inhibited by the bright surround; the target in shadow produces a low level of neural activity, but that level is only weakly inhibited. But again, this story works only if the exaggeration of differences is selectively applied only to reflectance edges, such as the edge between the target and its background. It has been shown (Gilchrist et al., 1983; Gilchrist & Jacobsen, 1984) that when the exaggeration is applied (as it must be by the structure-blind contrast idea) to the illumination edges within an image, the neural excitation of equal-lightness regions is made *more different, not less*.

These two examples show that the traditional question can be answered without really solving the problem. The traditional phrasing of the question is inadequate. A more adequate phrasing would be "How can changes in reflectance be distinguished from changes in illumination?" I suspect that had this been the standard form of the problem contrast theories as we know them would not have appeared. The important point is how the dominant metaphor shapes the question. It should be obvious that if luminance is coded at input, the problem to be solved is how lightness can remain constant when luminance varies. Alternatively, if luminance ratios are coded at input, the logical problem is one of distinguishing which ratios represent reflectance changes and which represent illumination changes. The definition of lightness constancy in the former terms rather than in the latter terms indirectly reveals something important about the guiding metaphor.

Wertheimer (1955) noted that, "It is conceivable, for instance, that a host of facts and problems have been concealed rather than illuminated by the prevailing scientific tradition" (p. 2). Perception of the illumination, disappearance of stabilized retinal images, and the constancy of lightness despite a change in its background are surely included in that host. I have tried to show that there has been a dominant photometer metaphor in the field of lightness perception, and that it has had real implications for the kinds of ideas that have been studied. The explicit rejection of this metaphor sets the Trieste group apart from most, but not all, others in this field. I believe this common element has allowed us to engage our differences in a more fruitful way.

4. MEDIATING PROCESSES FROM THE RELATIONAL PERSPECTIVE

Certain theoretical constructs are associated with the relational perspective in general, and the authors of these chapters in particular. Naturally these theoretical ideas are consistent with the relational conception of the input and the output, as discussed earlier. They struggle with the problem of how an input of luminance edges and gradients can be transformed into our familiar multidimensional experience of the world of surfaces. Not all these concepts are fully shared by each of the members of the group, and yet I believe the differences lie within a shared perspective.

4.1. Edge Integration

For quite some time after Wallach's classic ratio findings, the relational idea was mainly understood in the way Wallach presented it, as limited to a single luminance relationship between adjacent surfaces. It was only a matter of time before the problem of lightness constancy with respect to changes in the background (Type II) would come into focus. This and other problems have made it increasingly apparent that the visual system must have some way to compare nonadjacent regions of the image. About 1969, a number of publications appeared that assumed the visual system to be capable of extracting luminance ratios between remote, nonadjacent surfaces, by means of a mathematical integration of chains of local edge ratios. Whittle (Whittle & Challands, 1969) was one of the first to refer to such a process. However, Arend (Arend, Buehler, & Lockhead, 1971) produced a qualitative demonstration of the process in an elegant manner. Two equal luminance increments were placed within side-by-side background regions. The background regions were also identical (to each other) in luminance but separated by a Craik–O'Brien contour so that they appeared different. The layout was analogous to the classic simultaneous lightness contrast pattern found in textbooks except that the two backgrounds did not differ physically in reflectance; they only appeared different because of the illusory contour. It was an ingenious critical experiment, with three possible outcomes. If it is the apparent brightnesses of the two backgrounds that produce the traditional contrast effect, then the increment on the brighter appearing background should appear darker than the other increment. On the other hand, because the two increments are equal in luminance, and because their two backgrounds are also equal (physically that is), and because therefore the center/surround ratio of each increment is the same, there are ample reasons to predict that the two increments would appear the same. Only a process of edge integration would predict that the increment on the brighter appearing background would in fact appear brighter than the other increment, and that was in fact the result. Not long after this, Arend (1973) spelled out a mathematical model of edge integration as the complement of the differential processes of the input.

In its simplest form the idea is that the luminance ratio between two remote regions of an image is equal to the product of all the ratios (or gradients) that occur between the two regions. Land and McCann (1971) hypothesized a mechanism whereby the visual system integrates chains of ratios, each ratio being picked up by a very closely spaced pair of photoreceptors. This concept seems compatible with that employed by the present authors. Intuitive evidence for such a process comes from the observation that a gray surface can be viewed against a variety of backgrounds, differing greatly in luminance, with little loss of constancy in the gray surface, what I have referred to as Type II constancy, or constancy with respect to changes in background. A model of lightness without edge integration, based simply on the local edge ratio (such as a contrast theory) would require substantial changes in perceived lightness as the background luminance is varied, but edge integration allows the gray surface to remain in constant touch, so to speak, with a white standard even though the white standard is not adjacent.

Yarbus (1967, p. 96) had already published evidence for a process like edge integration by showing that, when targets of equal luminance are placed on side-by-side backgrounds of high and low luminance, respectively, and the border between the backgrounds is stabilized, the targets appear very different in lightness; the effect is far in excess of the usual contrast effect that occurs with that luminance pattern.

Piantanida and I conducted a very direct test of the edge integration idea in a stabilized image experiment (Piantanida & Gilchrist, in preparation). Our experiment, really a quantification of the Yarbus study (1967), was a logical sequel to Krauskopf's now-classic experiment (1963). It was a reversal of Krauskopf in which the outer boundary of the surround was stabilized but the boundary of the center was not. We reasoned that the center/surround relationship is treated by the visual system as a departure from the baseline defined by the outer border of the surround. This implies that, if the outer border of the surround is stabilized, not only should the perceived color of the surround change but the perceived color of the center should also change, *even though its border has not been changed.* Our experiment is illustrated in Fig. 1.1. We selected the conventional simultaneous lightness contrast pattern as the basis: equal gray squares on white and black backgrounds, respectively. We stabilized the entire closed boundaries of the white and black backgrounds but left the boundaries of the gray squares visible. The result was that the white and black backgrounds appeared as a homogeneous middle gray, and the square targets now appeared black and white, respectively. This result implies that the border dividing the white and black backgrounds is every bit as important to the lightness of the two targets as the border of each target itself, an implication fully consistent with the edge integration concept.

Gilchrist et al. (1983) showed that an analogous result can be produced, not by stabilizing the black/white border but by causing it to appear as an illumination edge. The simultaneous contrast display was embedded within a larger context that made the border between the black and white backgrounds appear as a sharp illumination border. This caused the targets to appear nearly black and nearly white, respectively. We argued that edge integration can explain lightness constancy if the visual system is able to exclude illumination edges from the integration.

This edge integration process seems to intrude sometimes where it is not wanted. I argued (Gilchrist, 1988) that this is the reason for the small loss of constancy in the Katz-type lightness constancy paradigm, which requires the observer to match the lightness of a pair of side-by-side color wheels, one standing in bright illumination and one standing in shadow. The illumination gradient dividing the shadowed background from the illuminated background is never completely excluded from the edge integration. Whittle (Whittle & Challands, 1969) argued that this is what produces the ambiguity in matching experiments using side-by-side center/surround displays (Hess & Pretori, 1894; Wallach, 1948). The task

1. THEORIES OF LIGHTNESS PERCEPTION

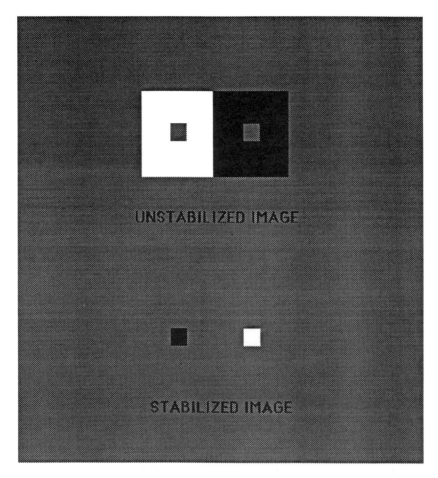

FIG. 1.1. Illustration of Piantanida and Gilchrist experiment. (A) Stimulus configuration: The outer borders of the black and white backgrounds were retinally stabilized, but the borders of the two gray squares were not. (B) Appearance of the display after stabilization.

produces a visual conflict. The targets can either be matched by matching their local center/surround ratios, or else they can be matched by producing an integrated ratio of 1. This is roughly equivalent to matching the targets for relative luminance or absolute luminance. Even when the context indicates that the former is required, the latter process is never fully excluded if conditions allow it. It is one of the many strengths of the Whittle & Challands paradigm that such an intrusion is prevented by their haploscopic technique. The visual system is prevented from finding an integrated ratio between the two targets because the borders of the two backgrounds are superimposed in the binocular image; they are not separately available for the integration process. As a result, the task in the Whittle & Challands experiment is easy to perform.

4.2. The Simplicity Principle

The principle of *prägnanz*, or simplicity, has long been considered an intuitively appealing idea. But it has always seemed to have a post-hoc quality. The Gestalt theorists were not able to provide an operational definition that would predict perceptual outcomes before the fact. Ironically, it is precisely the multiple dimensions of experience that make one of the thorniest problems for the minimum principle. Simplicity on one dimension generally has a cost in terms of complexity on another dimension. I believe the work of Gerbino and Bergström is very important in this respect. Each has been working to combine the phenomenological insights of the Gestalt school with greater mathematical rigor.

Gerbino is heir to a fine tradition of work on visual perception. He was trained by Kanizsa, well known for his emphasis on contour information in vision and in particular his striking demonstrations of subjective contours. Kanizsa was a student of Musatti, the great founder of Italian experimental psychology. Musatti's mentor Benussi, known for the Benussi ring, was a member of the Graz school. Musatti reinterpreted Benussi's work in terms of Gestalt psychology, closely associated with the relational perspective. Another of Musatti's students, Metelli, produced the most comprehensive analysis of perceived transparency. His "theoretical nephew" Gerbino has refined and extended that analysis.

Gerbino takes on this problem with an updated version of the minimum principle, one based on coding theory. According to this approach, the preferred perceptual organization is that one that entails the smallest number of parameters. For example, given an appropriate luminance pattern (see Fig. 5.1, chap. 5), the visual system would prefer to see a transparent layer in front of adjacent white and black surfaces rather than four adjacent shades of gray, because the code would require three (the black surface, the white surface, and the transparent layer) rather than four (one for each gray surface) values. The economy is even greater when illumination is factored in. The visual system strongly resists seeing a set of coplanar surfaces as if each has its own peculiar level of illumination. This may seem self-evident, but it is no less important theoretically.

Ironically, Gerbino showed that existing models have mistakenly assumed that the illumination on background is the same as that on the transparent layer. Empirical data reveal that these layers are treated as having different illumination values. Does this mean that the illumination is perceived as more complicated than necessary? Not according to Bergström, who postulates the principle that every scene is perceived as illuminated by the minimum number of light sources. For coplanar surfaces, this is the same as saying that the illumination level is the same on all surfaces. But for surfaces lying in different depth planes, as we have in transparency situations, it would be highly coincidental for the transparent layer and its background to have exactly the same illumination level. Here we see a close parallel between Gerbino and Bergström.

4.3. Analysis of Components

Bergström was trained by Johansson, who in turn studied with Katz, the noted phenomenologist of color appearance. Hence there is something deeply fitting about Bergström's effective application of Johansson's (1977) vector analysis model to "the world of color."

Bergström's is perhaps the most systematic account in this volume. In this view, a perceptual vector analysis renders retinal gradients into three separate representations of surface lightness, perceived illumination, and perceived three-dimensional form (see Fig. 6.5, p. 267). Adelson and Pentland (1990) recently provided an elegant illustration of this kind of scheme in their workshop metaphor. Like Gerbino's, Bergström's analysis is fundamentally based on a minimum principle. Analyzing reflected light into common and relative components is crucial to perceiving the complex pattern of luminance variations on the retina as caused by the fewest number of distal factors.

According to Bergström, reflected light is analyzed by the visual system into common and relative components. Both the common and the relative component depend in a fundamental way on luminance gradients. Although it is natural to think of the relative component in terms of gradients, this may not be entirely obvious for the common component. However, at the most basic level, to say that the perceived color of a homogeneous region is signaled by the change of light at its edge is to say that all points in the homogeneous region have something in common. (It is, of course, quite efficient that such a multitude of points can be represented by a single value—the edge ratio.) The edge defines the way in which all the points inside it differ from all points outside it. In this sense common and relative components are two sides of the same coin.

This approach can be seen in its barest outline in the familiar disk-annulus paradigm. In this simple paradigmatic display, it seems that both the common and the relative component are associated with a particular edge, the relative with the border of the disk and the common with the outer border of the annulus. Walraven (1976), in what could be called a chromatic version of the Whittle & Challands study (anticipated in Whittle, 1973), showed that the perceived color of a central disk is determined, under certain conditions, by only the difference in light between the disk and the annulus, a difference represented by the boundary between the disk and the annulus. The light that is common to the disk and the annulus, represented by the outer boundary of the annulus, does not contribute to the perceived color of the disk, at least not in the sense that its light is added to that of the overlain disk. The light common to both disk and annulus, like the common component in Johansson's motion analysis, serves as a frame of reference. The change in light at the disk–annulus border is seen as a departure from a baseline color associated with the change of light at the outer border of the annulus. Figure 1.2 may be helpful in grasping this. It illustrates Walraven's observation that what we have always regarded as a *disk surrounded by an annulus* is treated by the visual system as an *increment added to a pedestal* (or a decrement subtracted from one). The broader implication is that the retinal image is functionally decomposed into layers, rather than into adjacent color patches, as in a mosaic.

This fascinating relationship between relative luminance and layered components can also be seen in the Krauskopf (1963) experiment. He

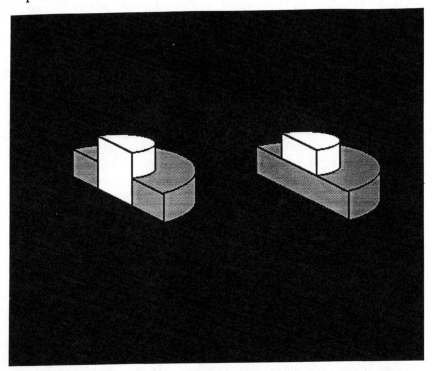

FIG. 1.2. (Left) Traditional mosaic conception of the disk–annulus configuration: A disk surrounded by an annulus. (Right) How it is treated by the visual system: An increment added to a pedestal (from Walraven, 1976).

demonstrated that, when the border between a colored disk and a differently colored annulus is retinally stabilized, not only does the disk disappear, as one would expect, but the color of the annulus appears also in the region where the disk had been, as if the disk had been filled in with the annulus color. *Filling-in* is the concept commonly used to understand this result, but I do not think it is the right one because it still implies a disk surrounded by an annulus. Because only the change in light between the disk and the annulus can strictly be said to have been stabilized, one would expect only *the layer of light defined by that change to disappear*. This should have no effect on the underlying layer of light defined by the change at the outer border of the annulus. That underlying layer should remain and indeed it does. We might say that when the relative component, signaled by the edge of the disk, is removed, the common component, signaled by the edge of the annulus, is revealed (it is already filled in).

The experiment described earlier that I conducted with Piantanida showed that when the outer border of the annulus is stabilized (whereas the border between each center and its annulus is not), both the center and the annulus change perceived color. The fact that the center changes perceptually even though the edge information at its border did not further shows that the filling in idea is too simplistic a concept.

Imagine a disk–annulus experiment with two disk–annulus patterns, side by side. The pattern on the left has the following luminances: annulus 4, disk 1; on the right side: annulus 40, disk 10. In fact, the observer will perceive each pattern as a middle gray disk surrounded by a white annulus, but with more illumination on the right-hand pattern. Here the lightness values are signaled by the disk–annulus luminance ratios in each display, but the illumination relationship is signaled by the outer boundary of each annulus. Again, the illumination component that is common within a given pattern becomes a relative component when the comparison is between the two patterns. There is a hierarchy of commonness.

4.4. Classification of Edges

Just as there are multiple dimensions within the perceived world, so the gradients in the retinal image have multiple causes. In fact, it could be said that every retinal gradient represents a change in one or another of the dimensions of the physical world that we perceive. A retinal gradient can be caused by a change in the reflectance of a surface, a change in the illumination on the surface (which in some cases is caused by a change in the orientation of the surface), the border of a transparent surface, the border of a veiling luminance, or the border of a glossy portion of a surface, to name some obvious examples (see Todd, 1985).

Somehow the visual system is able to distinguish these various kinds of gradients; if not, lightness perception would at times be wildly inaccurate. The problem of how the visual system is able to correctly classify retinal gradients remains a major obstacle, both for an adequate theory of human surface color perception, and for a workable machine model. Koffka (1935) anticipated this problem very early, and subsequent developments have not weakened the claim with which he accompanied his statement of the classification problem:

> Given two adjoining retinal areas of different stimulation, under what conditions will the corresponding parts of the behavioral (perceptual) field appear of different whiteness but equal brightness (or "illumination"), when of different brightness but equal whiteness? A complete answer to this question would probably supply the key to the complete theory of color perception in the broadest sense (p. 248).

This is the proper form of the lightness constancy question, as discussed on page 22. It implies a disentangling of illumination and surface reflectance, more than the discarding of the illumination. Gibson's (1966) question "Why is a change of surface reflectance not regularly confused with a change of illumination on the surface?" is an example of the insight that can be found in his surprisingly spare comments on lightness perception. We see here the influence of Koffka during their years together at Smith College.

The neglect of this question and the neglect of the multidimensional aspect of lightness perception in general can be understood, as noted earlier, by the dominance of the photometer metaphor and its point-wise approach. Even models that focus on edges (e.g., Cornsweet, 1970) have implicitly treated all edges as reflectance edges. The fact that illumination edges not only abound in retinal images but are correctly perceived as such has been illustrated by studying perception in environments where reflectance edges had been eliminated by painting every surface the same color (Gilchrist & Jacobsen, 1984). In fact, the various surfaces are perceived as being the same color, implying that all luminance gradients are attributed to changes in the illumination. Perhaps such an obvious result would be predicted by anyone who considers the situation. Yet this should not be allowed to detract from the theoretical impact. The retinal image in this case is highly articulated, with both sharp and shallow gradients, and a robust dynamic range. Essentially all existing models, taken at face value, would predict that a large range of surface colors (specifically gray shades) would be perceived. And yet the results show clearly that the variations in luminance are perceived as variations in local illumination, not surface lightness.

Recently a new generation of lightness and color constancy theories has arisen (Horn, 1974; Maloney & Wandell, 1986; Marr, 1982). They have been heavily influenced by the Land (1977; Land & McCann, 1971) retinex model, and they are closely associated with work on machine vision. They do represent certain advances over the older ratio and contrast models that take the center/surround pattern as paradigmatic. For one thing, the new paradigmatic display, the Mondrian, is a bit more complex than a simple center/surround display. Although the Mondrian, with its sharp-edged coplanar surfaces and limited luminance range, hardly captures the complexity of everyday life scenes, it is a step in the right direction. As for perceptual structure, at least the reflectance edge/illumination edge distinction is mentioned. However, it is hastily dismissed in a quite inadequate manner. According to this cliché, it is suggested that illumination edges tend to be gradual and reflectance edges tend to be sharp. Because the visual system is known to be less sensitive to gradual edges than to sharp ones, it is argued that the illumination edges drop out of the picture by falling below threshold for detection. Despite the emphasis on edge information, here again we find the same tired concept, earlier attributed to Helson, Hurvich and Jameson, and Hering, that visibility of illumination is sacrificed in the service of visibility of surface lightness.

There are a number of serious flaws in this story. First, many illumination edges are very sharp. Attached illumination edges are as sharp as the corners that produce them. Even cast illumination edges can be quite sharp, all the more so as the distance of the observer increases. Second, we do in fact perceive levels of illumination, both local and global, even when they are quite blurred. Relative edge sharpness probably does play a role in sorting out the two main types of edges, but it cannot be used in so facile a way to bypass the question of perceptual structure. Arend gives a good analysis of this issue (see p. 179), in particular giving a critique of the closely related claim that variations of illumination within the image are low spatial frequency whereas those of surface color are high.

In a well-known demonstration, Land and McCann (1971) showed that lightness constancy is good when an achromatic Mondrian is illuminated from below so as to cast a very gradual illumination gradient over the entire Mondrian. They posited a series of luminance comparisons of closely spaced pairs of points that, together with a minimum threshold luminance ratio, would be blind to the illumination gradient, picking up and integrating only the sharp reflectance borders. However, it can be inferred from an analogous experiment by Gilchrist et al. (1983) that Land and McCann would have obtained almost the same results had their illumination gradient been as sharp as the reflectance borders. Work by Arend & Goldstein (1989) confirmed this.

Bergström even argues (this volume) that it is precisely the ability to see Land and McCann's gradual illumination edge that allows the visual system to separate the illumination from the surface lightness, resulting in the achievement of constancy. The problem of perceptual structure, represented here by the problem of classifying edge types, will have to be confronted sooner or later.

Koffka's question has been a central one for much of my work. My work on lightness and depth perception (Gilchrist, 1977, 1979, 1980) showed that luminance gradients dividing coplanar surfaces tend to be seen as reflectance edges, whereas gradients dividing noncoplanar surfaces tend to be seen as illumination edges. In fact, illumination edges do divide coplanar surfaces in the case of cast shadows, but here is where edge sharpness would often be decisive; cast shadows normally have penumbrae. Yet there are cases where sharp illumination edges divide coplanar surfaces and perception is still veridical. As I showed in my experiments with Delman and Jacobsen (Gilchrist et al., 1983), correct edge classification is probably based on factors such as overall luminance range, figural (geometric) organization, and the presence of ratio-invariant intersections where illumination edges cross reflectance edges (difference-invariance is produced when two illumination edges cross one another).

Bergström (1977b) made a major contribution to the edge classification problem with his application of Johansson's vector analysis to lightness perception. Analyzing reflected light into common and relative components provides a way of splitting the retinal image into layers, much as a skilled stone mason, knowing just where to strike, splits a stone into its natural parts. Although many details have yet to be worked out, this kind of vector analysis offers a systematic approach to the long neglected but absolutely crucial problem of edge classification.

The authors of these chapters do not speak with a single voice. Bergström, Gerbino, and I would probably call ourselves gestalt theorists. Arend seems to be becoming a "Gibso–Helmholtzian." Whittle would probably reject all labels. Yet I argue that our work is mutually complementary and, taken as a whole, represents a major shift, one that is already well under way in the field, away from inadequate concepts and a too restricted definition of the basic phenomena.

ACKNOWLEDGMENTS

This chapter was supported by National Science Foundation grants BNS-8909182 and DBS-9222104.

2
The Psychophysics of Contrast Brightness
Paul Whittle
University of Cambridge

1. CONTRAST BRIGHTNESS

1.1. Introduction

The perception of color must depend on both sensory and cognitive processes. This chapter and the next put forward two main arguments. First, that there is a simple and coherent psychophysics of contrast colors, which includes the phenomena of simultaneous contrast and adaptation and reflects the contrast coding that we know to be a prime function of early vision. Second, that the higher levels of the visual system must, because of the variety of circumstances under which we can see, use the contrast-coded signal in radically different ways in different contexts. The higher processes can mask the lower, so that to understand the system we must distinguish these different levels.

Most of my discussion concerns the intensive dimensions of color: lightness and brightness. The observation that brightness depends on relative luminance, and the hypothesis that this is mediated by retinal interactions, go back at least to Mach (1865).[1] One variant is the appeal-

---

[1] We need a few definitions. *Brightness* is apparent luminance: the apparent intensity of the light reaching you from any part of the visual field. *Lightness* is the apparent reflectance of surfaces. These troublesome concepts are discussed further in the next chapter. I describe brightness as *contrast brightness* in those situations in which it is primarily determined by local luminance contrast. *Increments* and *decrements* are light added to or taken away from a background, respectively. *Simultaneous brightness contrast* or just *brightness contrast*, or *brightness induction*, is the subjective phenomenon in which the brightness of a light is inversely

ingly simple idea that the luminances of a uniform region and its surround affect brightness only insofar as they affect the contrast at its edge. Wallach's (1948) often quoted "ratio rule" ("... perceived colors ... depend on the ratios of the intensities of the reflected light") suggests this. A persuasive dynamic demonstration is as follows. Set up two disk–annulus displays, side by side. Set all four luminances to the same medium value. Modulate the luminances of one disk and the opposite annulus together but 180° out of phase, so that when one is maximum the other is minimum. (Use a computer monitor or suitably oriented pieces of polaroid and neutral filter for a projected display.) Then both disks alternate between bright and dark, and observers find it difficult at first to tell which is being directly modulated, and which only "by induction." This shows that contrast, the relative luminance of region and surround, is the crucial variable.

However, although the dependence of brightness on contrast is an old idea, and although models based on receptive field characteristics have had considerable success in explaining Mach bands and similar phenomena, ordinary simultaneous contrast of the kind just described remains surprisingly controversial (Fiorentini, Baumgartner, Magnussen, Schiller, & Thomas, 1990). One problem is that uniform regions look uniform. Here is another. If brightness depends simply on contrast, then increments, which have positive contrast, should always be brighter than decrements, which have negative. But this was not true in the classical measurements of Hess and Pretori (1894), and has rarely been shown in the many studies since then. My argument in this chapter is that it is true in some situations, and not just as an isolated finding, but as part of a coherent pattern of results deriving from several different techniques. I describe studies of brightness by the methods of matching, scaling, and discriminating that converge on several common conclusions.[2] The agreement is close enough for the outlines to appear of a simple quantitative model to accommodate all the different kinds of data. Further support is provided by experiments in which the luminance of light or surround is modulated in time, as in the preceding demonstration, and by parallel results in the hue domain.

In those situations in which brightness depends primarily on contrast, we can regard brightness and lightness as forming a single signed dimen-

---

related to the luminance of a neighboring light. I use it here to describe the brightness of uniform regions (usually in uniform surrounds); this is sometimes called *area contrast*. The word *contrast* by itself in these chapters always refers to *physical* contrast, that is, relative luminance.

[2]There is a considerable amount of detailed psychophysics in this chapter. Those who wish to avoid such minutiae can get the gist from sections 1, 2.1, 4, 6.1, 9.4, and 10. Some of this psychophysics has not been published before, and none has been presented in an integrated review that shows the relations between the different studies.

sion that I call *contrast brightness* to emphasize the relationship to contrast and to distinguish it from brightness in other situations. The positive part is the brightness of increments, which look white or self-luminous, and the negative part the lightness of decrements, which look like gray surfaces or black holes. In between we have zero brightness, or invisibility, corresponding to zero contrast. Many aspects of contrast brightness accord with current knowledge of retinal function. This is the domain of bottom-up models of brightness perception such as those of Hamada (1984) and Kingdom and Moulden (1991), but the emphasis here is on the brightness of uniform regions rather than on border-contrast effects.

In the next chapter I show that in other situations lightness and brightness can be almost independent of contrast. To understand this requires going beyond the psychophysics of early vision and considering the problems posed by the perception of surfaces and lighting in natural scenes. We need to consider constancy of surface color with respect to changes both in illumination and in the color of neighboring surfaces. The dependence of surface color on local contrast can contribute to the first kind of constancy but would prevent the second. Therefore, our accurate perception of surface color under variations of both illumination and background surfaces implies that high level (cognitive) processes can use the contrast signal in a way that is sensitive to the overall parsing of the scene into illumination and surface colors. Failure to distinguish this level from the earlier one has led to persistent confusion in the literature on brightness and lightness, in spite of the fact that from Helmholtz to Marr there have been many discussions of the relative roles of sensory and cognitive processes on the perception of color (e.g., Beck, 1972; Evans, 1948; Helmholtz, 1866; Marr, 1982; Whittle & Challands, 1969; Woodworth, 1938).

Contrast brightness may strike some readers as a somewhat slippery concept, by comparison with formally defined photometric quantities. It is not intended to be defined with formal precision. It is brightness in those situations in which brightness is most strongly determined by local luminance contrast. Its function is to act as a label for a group of phenomena that occur together to some extent. The concept is intended to emerge from the discussion in these chapters, rather than being clearly definable at the start. It emerged for me while I was writing and had a surprisingly clarifying effect, though whether I have managed to communicate that is another question. In many ways it is analogous to *stereo depth*, both in the work the concept does and in its specific properties (Brookes & Stevens, 1989). Stereo depth is apparent depth in those situations in which it is most strongly determined by binocular disparity. It can be thought of as a bipolar relative dimension. It has certain characteristics, such as being unequivocally "in front" or "behind," and hyperacuity around the reference level, that usually go together. It requires special situations to demonstrate it in a pure form. In ordinary vision it is overlaid and can even be reversed by other depth cues. In all these respects, it is analogous to contrast brightness.

This analogy suggests a wider perspective. Contrast coding is probably very general (section 6.5). We can define it as transforming a stimulus dimension I into a signal of magnitude $\Delta I / f(I)$, where ΔI is a difference with respect to some reference level, and $f(I)$ is a function of the absolute level. Such a code, combining differencing with multiplicative attenuation—normalization—has many advantages, some of which are exemplified later. However, its two components pose two complementary problems whenever we need to recover information about absolute intensity: What is the zero, and what is the scale? Examples in the context of lightness are the anchoring problem (what is white in this scene?), and the scaling problem (what is the range of grays?).

1.2. A Little History

This work started as a study of simultaneous brightness contrast. The phenomenon has been known for centuries, and frequently measured. Jung (1973) told us that it was "first clearly formulated by Leonardo da Vinci in his *Treatise on Painting* written around 1500" (p. 26). One has to be an optimist to think that more measurements will enable us to understand it better. However, three discoveries in the last 50 years have stimulated just that hope. The first was the discovery of lateral inhibition in the retina (Hartline, Kuffler, Barlow). This produced a cluster of new studies and discussions of brightness contrast centered on the idea of lateral inhibition (see Brown & Mueller, 1965; Cornsweet, 1970). The second was the demonstration of the fading of stabilized images (Ditchburn, Riggs, Yarbus), and the third, Hubel and Wiesel's finding that visual cortical neurons responded only to edges or other spatial transients.

I began after the third of these, when edges and contrast were on people's minds. Working on binocular rivalry, that theme park of edge effects, had impressed on me the importance of the sign and magnitude of contrast. Increments in one eye fuse with congruent increments in the other, but rival with decrements, and the proportion of time for which you see a rivalling stimulus increases with contrast (Whittle, 1965). As I manipulated patch and surround luminances, I was impressed also by the dramatic brightness changes, and it seemed to me that they too depended on the sign and magnitude of contrast in a mathematically simple way that was missed by lateral inhibition models of brightness contrast (although caught by Wallach). They missed it because they were not conceived in terms of edges or contrast at all. Gilchrist, who describes in chapter 1 how the same phenomena captured his interest, would say they were trapped in the "photometer metaphor." I pursued this hunch intermittently in a series of studies over the next 30 years.[3] This chapter attempts to summarize them and to see how much the hunch was justified.

2. MATCHING CONTRAST BRIGHTNESS

2.1. Outline of Experiments and Data

Much visual psychophysics is concerned with detection thresholds—the almost invisible—and when it does study things we can see clearly it tends to favor performance measures, such as discrimination or latency, over judgments of appearance. A hope behind the present work, on the other hand, is that a psychophysics of appearance is possible that extends the findings of threshold psychophysics, with little loss of rigor, into the suprathreshold domain. In trying to realize this hope, I have found that a useful rule is to use, when possible, experimental situations in which the judgments required from observers are subjectively satisfactory in the sense that they seem to be compelled by the stimuli, rather than feeling like "matters of opinion." It is my experience that when tasks are selected with this criterion in mind, the results are more likely to be both reliable and comprehensible.[4]

Whittle and Challands (1969) measured thresholds and matched the suprathreshold brightnesses of the two square patches in the display shown in Fig. 2.1a. Two features are important. First, each eye saw a *center/ surround* display: a uniform patch of light in a larger uniform surround. This is the logical choice of stimulus to study the dependence of brightness on edge contrast. It has of course been used very extensively in vision research, and it can be argued on several grounds that it should be regarded as a fundamental stimulus pattern for analyzing vision. This is particularly so for studying color appearance, as has been pointed out by,

---

[3]The experiments were begun in Lorrin Riggs's stabilized image laboratory at Brown, after working on rivalry in Cambridge (England) alongside Peter Burgh and "C" Grindley, who were studying brightness contrast.

[4]This may sound obvious, but it is often disregarded. Even in the deliberately simplified conditions used for most psychophysics, some tasks feel "natural," that is, easy and/or subjectively satisfactory in the sense described. Others feel forced. That is particularly likely when they are constructed intellectually to answer a preconceived question about mechanisms. My suggestion is that tasks that feel natural are more likely to be using our sensory and perceptual capacities in a natural way and therefore are more likely to tell us something important.

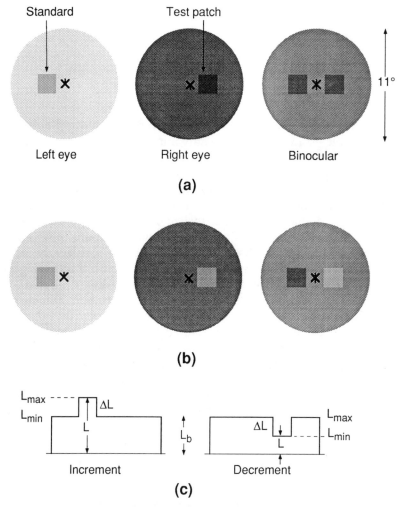

FIG. 2.1. (a) and (b) The haploscopic superimposed-background display (HSD). The sides of the squares, and the gap between them, are about 1°. To make brightness matches the subject adjusted the value of ΔL for the test patch, seen by the right eye, on various background luminances, L_b, to match in brightness the standard seen by the left eye. Subjects always match decrements to decrements, as in (a), or increments to increments, but never increments to decrements as in (b). (c) The notation for luminance.

for instance, Evans (1974) who called it the *simplest general stimulus*, and Mausfeld and Niederée (1993) who called it the *minimal relational stimulus*.

The second important feature of the display, which is more unusual, is that the left and right eye surrounds are binocularly fused so that the two patches, although seen by separate eyes, appear to be on the same background. This arrangement, which was introduced by Hering (1890) to demonstrate color contrast and used by many others since then, is referred to often enough to deserve a special name: the Haploscopically Superimposed Display (HSD). It turns out to have three advantages over the displays traditionally used for studying brightness contrast, in which the surrounds are seen side by side. These advantages are all ways in which the HSD simplifies the measurement of the effect of a background on brightness. First, it makes setting a brightness match between patches on backgrounds of different luminance subjectively easy. The measurements cry out to be made: It is almost like doing visual photometry. In side by side displays the task is much more difficult, and subjects often feel there is no satisfactory match (section 3.6 of chap. 3). Second, an encouraging qualitative result was clear from the start: Increments were never matched to decrements, because any increment looked brighter than any decrement, whatever the four luminances involved.[5] This goes with the brightness contrast effect in the display being very large, perhaps the largest possible. The separation of increments and decrements is what you would expect if brightness was a simple function of the physical contrast between patch and background.

The third advantage appeared in the quantitative results, shown as families of isobrightness curves in Fig. 2.2. The upper set is for increments and the lower for decrements. The ordinate, LogΔL, increases upward for increments and downward for decrements, to give the graph a symmetry reflecting that of the stimuli.[6] The abscissa is background luminance.

---

[5]The reader may be able to test this informally with Fig. 2.1b. The "standard" and "test" gray squares for the left and right eye may not look very different, but if you cross your eyes and fuse the two displays, the squares look as clearly different as in the "Binocular" section of the figure. Note further that if while viewing the fused display you close one eye the remaining square changes little if at all in appearance. Its appearance depends only on the luminances in the viewing eye.

[6]The luminance notation is shown in Fig. 2.1. The more important definitions are grouped together here for reference. L is the luminance of the test patches, L_b that of background, L_{max} and L_{min} are the larger and smaller, respectively, of L and L_b. $\Delta L = L_{max} - L_{min}$. Luminance will usually be expressed as troland values—luminance times pupil area—but it is clearer to refer to it simply as luminance. t denotes flash duration. *Physical contrast*, the relative luminance of two regions, is commonly expressed as *Michelson (or Rayleigh) contrast* = $M = (L_{max} - L_{min})/(L_{max} + L_{min})$. We also use the *Weber fraction* = $\Delta L / L_b$ and $W = \Delta L / L_{min}$ and other *contrast expressions* of the form $\Delta L / f(L, L_b)$ where $f(L, L_b)$ is some function of L and/or L_b.

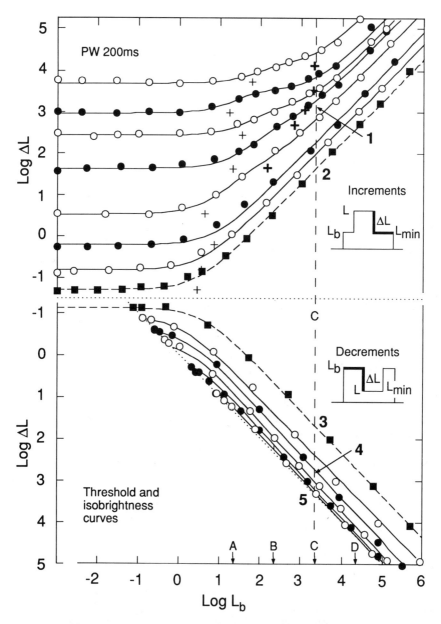

FIG. 2.2. Threshold (square symbols) and isobrightness (circles) curves for cone vision. Each curve is the average of between 2 and 10 runs in different experiments over a period of several years. The light was "white," deriving from a tungsten quartz-iodine bulb plus neutral filters. The units are trolands in this and all subsequent figures except where otherwise stated. The curves follow Equations 2 and 3. The numbers on line C correspond to those in Fig. 2.14.

2. THE PSYCHOPHYSICS OF CONTRAST BRIGHTNESS   43

Attend for now just to the increment data.[7] The square symbols and dashed curve show detection thresholds. The higher curves are suprathreshold isobrightness contours. The points along each represent matches to the same standard brightness. This is the parameter: the *brightness* of the left-eye standard.[8] The similarity of threshold and suprathreshold data, which is clear in Fig. 2.2, is the third advantage of the HSD. It promised to reduce two mysteries to one: to explain both suprathreshold brightness, including brightness contrast, and difference threshold in terms of the same process, light adaptation.[9] It was intuitively appealing that the threshold curves were members of the family of isobrightness contours because they are the bounds of the region of invisibility (zero contrast brightness) between positive increments and negative decrements and so might be expected to have constant "just noticeable" brightness. But this was also surprising, because it implied complete *discounting* of the background brightness, to use the term introduced by Walraven (1976; section 8.3).

Why the HSD has these advantages is discussed in the next chapter. In this one, I describe and analyze the data obtained with it. First, I discuss the pattern of isobrightness curves in more detail.

---

[7]Only increments were dealt with in the 1969 paper. Decrements are discussed later in the chapter, but the data are presented together in the same figures, where possible, to facilitate their comparison.

I have obtained families of curves similar to those in Fig. 2.2, though usually less complete, for several other subjects. Many of the data in this chapter are for myself as subject. The original papers give data for additional subjects, which in all cases, unless explicitly stated, support the conclusions I draw here. I regret seeming to put so much emphasis on my own data, but because I am making interstudy comparisons, and I am often the only subject in common between studies, it seems the best procedure.

A feature of the main measurements that should be pointed out, although I provide evidence later that it is less important than might at first be thought, is that the patches were presented as 200ms flashes added to or subtracted from otherwise uniform backgrounds. 200ms is about the duration of a fixation in normal vision. A flash of this length usually has a single unambiguous brightness. With longer durations brightness changes during the exposure, so there is ambiguity as to which part to base the judgment on. We justified the use of both flashes and the center-surround display in terms of the idea that perception depends on transients. These features amount to giving the stimulus the same bar profile in both space and time: an on-step followed by an off-step (or vice versa).

[8]The same isobrightness contour can be produced by very different left-eye standard *patch and surround luminances*, provided they are chosen to give the same patch *brightness* by matching it to the same right-eye patch. This amounts to claiming that brightness matching in this situation is transitive (Whittle & Challands, 1969), which is crucial for the validity of the method.

[9]The identity of Wallach's "Ratio Rule" to Weber's Law, curiously unemphasized by Wallach, had already pointed to this. However, Wallach used only decrements and a narrow range of luminances.

2.2. An Equation and an Ideal von Kries Scheme

The shape of the curve relating threshold to background luminance, which Stiles called a *threshold versus intensity* (t.v.i.) function (Wyszecki & Stiles, 1982), is approximately described by the equation

$$\frac{\Delta L}{(L_b + L_0)^n} = c \tag{1}$$

where L_0, n, and c are constants (Fig. 2.3a). On a log–log plot n is the slope of the upper part of the curve. For the increment data in Fig. 2.2, and in

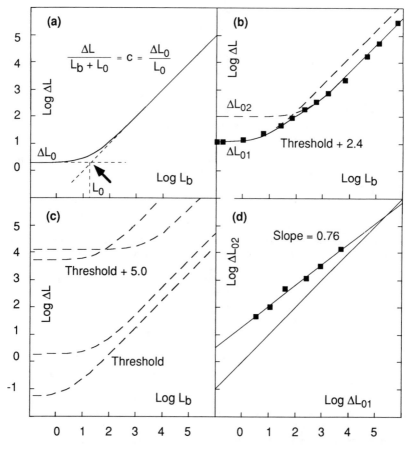

FIG. 2.3. The ingredients of the quantitative description of increment isobrightness curves. (a) A t.v.i. curve following Equation 1. (b) A two-branched fit of two t.v.i. curves to data points like the upper sets in Fig. 2.2. (c) The relative positions of the two branches at threshold and 5.0 log units above threshold. (d) The plateau height for the upper branch plotted against that for the lower.

Fig. 2.3(a), $n = 1$ (Weber's Law) as it generally is for increments in cone vision when the test patch is more than a few minutes in width. L_0, sometimes interpreted as noise or "dark light," produces the transition between the upper part and the horizontal plateau at low backgrounds. The position of such a curve is conveniently defined by the point where these two linear asymptotes meet, arrowed in Fig. 2.3(a). This has coordinates $(L_0, \Delta L_0)$, where ΔL_0 is the value of ΔL when $L_b = 0$, that is, the absolute threshold or the height of the low-luminance plateau.

Equation 1 suggests that brightness is constant if and only if the contrast expression $\Delta L / (L_b + L_0)^n$ is constant. This claim, which will be made more precise, is the main justification for the term *contrast brightness*.

The pattern of isobrightness curves in Fig. 2.2 suggests an ideal scheme in which all the suprathreshold curves have this same shape, differing from the threshold curve only in their vertical position on the graph. Equation 1 would describe all of them, with c as the only parameter changing. A given background luminance would attenuate all inputs equally, whatever their strength. The notion that adaptation produces such a multiplicative gain change is familiar from the Coefficient Law proposed by von Kries (1904) to describe chromatic adaptation. In Equation 1, however, the attenuation applies to the luminance difference, ΔL, rather than to the absolute intensity as in von Kries' rule.

This scheme expresses such a simple model of adaptation that it makes a useful reference frame even though the actual data deviate from it. Much of the discussion in the next sections is concerned with the deviations. This organizes the discussion but is not meant to imply that the ideal scheme is in some sense the fundamental truth, of which actual measurements are distortions.

2.3. The Brightness of Increments

2.3.1. *Increased Exponential Gain on Bright Backgrounds.*

For increments the most obvious deviation from the ideal scheme is the convergence of the curves toward the upper right corner of Fig. 2.2. The rightward drift of the upper curves makes their Weber fractions less than they would be in the ideal scheme. We can describe this as an increased sensitivity to the test light on bright backgrounds, or a decreased sensitivity to the background. The commonest description has been as an increased "gain" of brightness with respect to LogΔL on brighter backgrounds (Chen, MacLeod, & Stockman, 1987; Whittle & Challands, 1969). This is a different sense of gain from that in the previous section. The term *gain* is used somewhat loosely both in systems theory and in vision science to mean a multiplying factor, or the slope of the input–output function, at some stage in the system. Gain in Equation 1 refers to multiplying ΔL by the factor $1/(L_b + L_0)^n$, which *decreases* with L_b and hence is often described as *attenuation*. But in Fig. 2.2 the ordinate is the logarithm of ΔL, so if the convergence of the curves is described as a gain change, it is a gain that multiplies $\text{Log}\Delta L$. This is equivalent to raising ΔL to some power ($p\text{Log}\Delta L = \text{Log}(\Delta L^p)$) so I call this *exponential gain*. For increments this gain *increases* with L_b, so the two gains must be to some degree independent. The exponential gain will not be explicitly represented as an exponent in the equations I use. For instance, in Equation 1, it depends on how L_0 and c vary with brightness level.

The next sections consider five possible influences on the convergence of the increment curves: intrusion of different π-mechanisms, changes in stimulus duration or area, scattered light, and adaptation to the test light. Along the way Equation 1 will be modified to fit the data better, to make it a more useful tool for the subsequent analyses.

2.3.2. The Two-Branched Shape of Higher Isobrightness Contours.
In Fig. 2.2 the higher isobrightness curves change shape as well as shifting to the right relative to the lower ones. Their form suggests branches for separate π-mechanisms as in Stiles' two-color threshold experiments (Wyszecki & Stiles, 1982, p. 529; Fig. 2.3(b) illustrates). This interpretation was supported by the experiments of Whittle (1973), which were like those just described except that the stimuli were not white but were red or green increments on red or green backgrounds, in all four possible combinations, and the flash duration was reduced from 200ms to 40ms. When the color of the background was changed, the two branches drawn through each set of data points shifted horizontally by different amounts. This showed that the branches had different spectral sensitivities to the adapting field. When the flashes were green, the magnitude of the shifts agreed with the published field spectral sensitivities for π_4 and π_4'. These are Stiles' middle-wavelength mechanism in its normal and modified high-intensity condition. The data for red test flashes followed Stiles' long-wavelength mechanisms. The fact that suprathreshold brightness followed the same two-branched scheme as threshold suggests that brightness under these conditions is determined by the most active π mechanism, just as threshold is.

2.3.3. One Equation for the Increment Data.
The two-branched scheme also fits the data for white light in Fig. 2.2. Figure 2.3b shows how a typical set of isobrightness points can be well fit by two branches with the form of Equation 1. Suppose the lower branch has equation $\Delta L = c_1(L_b + L_{01})$ and the upper, $\Delta L = c_2(L_b + L_{02})$. Then the equation

$$\frac{1}{\Delta L^2} = \frac{1}{\left(c_1(L_b + L_{01})\right)^2} + \frac{1}{\left(c_2(L_b + L_{02})\right)^2} \tag{2}$$

combines them according to the rule that threshold or brightness is that of the most active mechanism, except that it allows some summation when the sensitivities are nearly equal.[10] The solid line in Fig. 2.3b follows this equation and the dashed lines show the separate branches where they are distinct.

It is useful to express all the coefficients of Equation 2 as functions of a single parameter representing brightness level. An obvious parameter to use is the plateau height of the lower branch, ΔL_{01}, which is the luminance required to produce that particular brightness on a dark background. The values of the parameters (c_1, L_{01}, c_2, L_{02}) that gave the best fit to each data set in Fig. 2.2 were estimated. They were used to generate the curves drawn in that figure, which all follow Equation 2. The +'s show the points (L_{01}, ΔL_{01}) and (L_{02}, ΔL_{02}) corresponding to the arrowed point in Fig. 2.3a for the upper and lower branches. Regression lines fit to the crosses gave $\text{Log} c_1$, $\text{Log} L_{01}$, and so on as linear functions of Log ΔL_{01} to a good approximation. Equation 2 could then be expressed in the form $\Delta L = f(L_b, \Delta L_{01})$, describing the shape of an isobrightness curve with ΔL_{01} as the only variable parameter. In Figs. 2.7 and 2.8 following, all the curves, one- and two-branched, through the increment data (the same data points as in Fig. 2.2), follow this single-parameter form of Equation 2.[11] It fits this large set of data quite well. (And also some that are not shown, and, with different coefficients, data for a second subject.)

Figure 2.3c and 2.3d show how it is that Equation 2 can describe all the suprathreshold curves whether or not they show an upper branch. Figure 2.3c shows the positions of the two branches at threshold and 5 log units above it. At threshold the upper branch is entirely above the lower, that is, it represents a mechanism that is less responsive whatever the background luminance and so does not influence the outcome. However, when the flashes are bright, this mechanism is more responsive on brighter backgrounds because its (exponential) gain is higher and it shifts to the right more than the lower branch. The higher gain is shown in Fig. 2.3d in which $\text{Log}\Delta L_0$ for the upper branch is plotted against that for the lower: The slope is much less than 1.0.

The single-parameter form of Equation 2, which we can call an

---

[10]This "root sum inverse square" trick combines equations describing two threshold mechanisms to yield a single equation that provides a lower envelope of the two mechanisms (i.e., threshold is that of the more sensitive mechanism, except where the sensitivities are nearly equal, where it approximates probability summation). Suprathreshold, the interpretation is that the response is that of the more active mechanism. This trick has often been used in psychophysics, for example, in line-element models.

[11]The actual equation, with parameter values for PW, was $1/\Delta L^2 = 1/(0.166 * \Delta L_{01}^{0.757} * (L_b + 6.026 * \Delta L_{01}^{0.243}))^2 + 1/(0.163 * \Delta L_{01}^{0.382} * (L_b + 111.7 * \Delta L_{01}^{0.382}))^2$. The log–log regression lines for the parameters (which accounted for 89% of their variance, on average) are here written as power functions.

isobrightness equation, is a useful computational tool. Given any increment on any background, we can now calculate what increment would look the same brightness on any other background. Although the equation was based on matches made only in one particular situation, for many purposes this is a less serious restriction than might at first be thought. In particular, the values of stimulus area and duration that were used seem to be representative of a wide range of values. Note that the computational use of the isobrightness equation does not depend at all on the validity of its interpretation in terms of π mechanisms. For computational purposes, all that matters is that the equation fits the data; an uninterpretable polynomial would do just as well.

2.3.4. *With Short Flashes, Each π Mechanism Obeys the Ideal Scheme.* So much for the changing shape of the higher isobrightness curves. In Fig. 2.2 they also shift to the right at higher intensities, and this is true for both upper and lower branches. However, it was not true for the experiments with colored flashes of Whittle (1973). There each separate branch moved vertically as brightness was increased. The ideal von Kries scheme did apply when brightness was predominantly determined by a single π mechanism.

Similar behavior was shown by rods, in an experiment designed to isolate them. Brightness matches were made between two 50ms flashes presented 11° below the fixation point as shown in the inset to Fig. 2.4. The subjects were dark-adapted for 30 min before the experiment, and the methods of Aguilar and Stiles (1954) were used to ensure that the flashes were seen with rods rather than cones over as wide a range as possible. The test flashes were green and the background red. The test flash entered at the edge of the dilated pupil. This favors rod response because cone sensitivity falls off more with oblique incidence than does rod sensitivity (the Stiles–Crawford effect; see Wyszecki & Stiles, 1967). Two procedures were used to determine whether the flashes were seen with rods or cones. First, the cone thresholds were measured by taking dark adaptation curves in the presence of different background luminances and finding the heights of the cone plateaux. These are shown by the dashed line and X's in Fig. 2.4. Second, on four of the brighter backgrounds, marked a to d in Fig. 2.4, settings were made with both central and peripheral pupil entry. The difference is shown by the arrows, with the tip representing the setting made with central entry. The length of the arrow therefore shows the size of the Stiles–Crawford effect and should be small if the response depended primarily on rods. This can be seen to be the case below cone threshold but not above. (For the points on backgrounds a to d where an arrow cannot be seen, it is smaller than the plotting symbol.) Subjects' descriptions of the appearance of the flashes confirmed the other indices of cone activity.

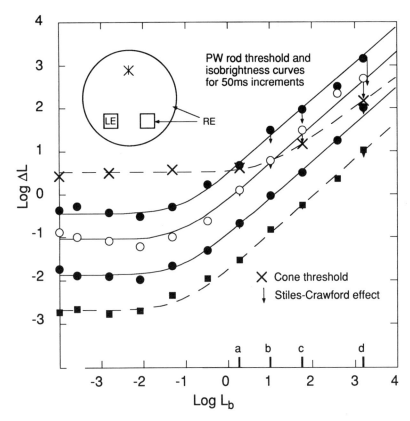

FIG. 2.4. Threshold (square symbols) and isobrightness (circles) curves for rod vision. (Unpublished data.) Test patches 1.9° square, 2.4° apart, 11.3° below fixation point; duration 50ms. Left-eye standard: dark background, green test patch. Right-eye: red background, green test patch entering dilated pupil 2mm below center. Xs show cone plateau height from dark adaptation curves. Units are photopic trolands. A second subject gave similar results.

Figure 2.4 shows the resulting threshold and isobrightness curves. The lines all follow the equation $\Delta L/(L_b + 0.087)^{0.86} = c$.[12] They are all the same shape, and are displaced only vertically as c varies. Thus the rods also, under these conditions, obey the ideal von Kries scheme. Because the two-color data for cones depended on the difficult procedure of strongly heterochromatic brightness matching, it is useful to have this confirmation from scotopic experiments where that particular problem was absent over most of the range. Note that in the upper right-hand part of the figure,

---

[12]Sharpe, Fach, Nordby, and Stockman (1989) found a slope constant of 0.77 for similar rod-isolation conditions, comparable to the 0.86 in this equation.

where cones are influencing the match, the isobrightness curves seem to continue their course unchanged up into the mesopic region. We return to this point in the next chapter.

2.3.5. *Duration and Brightness.* How much is this ideal von Kries behavior due to restricting vision largely to a single π mechanism and how much to using shorter flashes? What happens to photopic white-light measurements like Fig. 2.2 if duration is changed? We can get a partial answer to this from a variant of the brightness matching experiment in which flash duration, t, was changed during a run rather than L_b. The standard left-eye flash lasted 200ms. The right-eye flash varied in duration, but the two flashes always ended together. The subject adjusted the right-eye luminance so that the flashes matched at the end of their exposures. The condition of ending together made for easier matching than other temporal relationships, but even so when the durations were very different, it was a difficult judgment, not subjectively satisfactory in the sense discussed previously. However, there was reasonable qualitative agreement, both between the two subjects and with respect to similar published experiments.

Figure 2.5 shows some results. As in Fig. 2.2, the lowest curve is threshold and the higher curves isobrightness loci, but the abscissa is now Log duration. The left panel shows measurements on a dark background, and the others on backgrounds of 1.06 and 4.18 Log trolands. These data follow a familiar pattern. The left-hand part of each curve has a slope of –1.0, representing Bloch's Law for energy integration over time. For durations in this range it is the total energy in the flash, $\Delta L*t$, rather than luminance, that determines brightness. Note that it is ΔL that is integrated, not L. The right-hand part of the suprathreshold curves, of positive slope, represents the Broca–Sulzer effect: Brightness decreases with duration beyond about 100ms. In isobrightness curves this effect becomes infinite at high flash luminances because intense flashes are brighter than any steady light (Craik, 1940).

From data like these I estimated how the isobrightness curves in Fig. 2.2 would change if different flash durations were used. For durations of 40ms or less, which are in the Bloch's Law range at all brightness levels, the shift to the right of both branches of the upper curves was reduced.[13] However, it was not abolished, as it was for the colored flashes. This was shown more simply by making brightness matches with background luminance, rather than standard brightness, as parameter. If LogΔL on a bright background is plotted against the matching LogΔL on a dim one, the two L_b values being

---

[13]Reducing t below 40ms shifts all the curves upward without changing their relative positions. This follows from Bloch's Law.

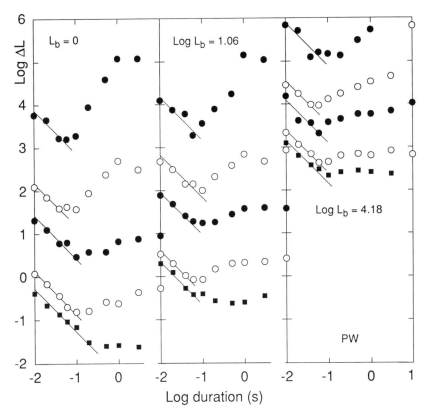

FIG. 2.5. Threshold and isobrightness contours for changing flash duration. (Unpublished data.) The display was as in Fig. 2.1, and fixation was steady. As in Fig. 2.2 the lowest curve is threshold, and the higher ones isobrightness curves. The parameter within a panel is the brightness of the left-eye standard, and between panels the luminance of the right-eye background. The straight lines have slope −1, representing complete temporal integration.

chosen so that brightnesses are determined by the same branch, a line with slope <1 results (like Fig. 2.3d, though that compares different branches). The slope is the ratio of the exponential gains at the two L_b values. In the ideal von Kries scheme, with no rightward shift, it would be 1.0. The slope increases as flash duration is decreased but does not reach 1.0.

If the duration is made longer than 200ms, the shift to the right is increased. However, up to moderate brightnesses, the change is sufficiently small that measurements with steady lights yield almost the same pattern as those with 200ms flashes (Fig. 3.1 in next chapter). This is important, because it allows us to take Fig. 2.2 as a description of simultaneous brightness contrast, a term that is generally taken to apply to steady lights.

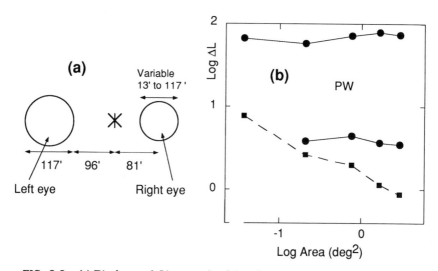

FIG. 2.6. (a) Display, and (b) a sample of data from an experiment in which the brightness of a patch of variable size was matched to one of constant size. (b) shows threshold (square symbols) and two isobrightness contours (circles). The corresponding left-eye standards were about 1.0 and 2.5 log units above threshold. Both test patches were seen against circular backgrounds of diameter 11° centered on the fixation cross, haploscopically fused in the HSD. The luminance of the background was 0.93 log td in the left eye and 1.96 in the right. Flash duration was 200ms. A second subject produced similar results. (Unpublished data.)

2.3.6. *Area and Brightness.* The pattern of isobrightness contours in Fig. 2.2 is not critically dependent on the area of the test patches. Figure 2.6b shows some data in support of this claim, from an experiment in which brightness matches were made between patches of different size. The isobrightness contours are flat: Area has no effect over this range. Diamond (1962) reached the same conclusion for a range of patch diameters from 3' to 27'.[14]

Note, however, that threshold depends markedly on area over the same range, as is well known (square symbols and dashed line in Fig. 2.6b). A related result was obtained for the apparent contrast of gratings by, among others, Georgeson and Sullivan (1975): Contrast sensitivity functions are much flatter suprathreshold than at threshold. So in this family, unlike the others, the threshold curve is not an isobrightness contour: Small flashes look much brighter at threshold than do large ones. Other data showed

---

[14]There are, however, some contradictory data in the literature. Diamond reported some from other workers, and Ogawa, Kozaki, Takano, and Okayama (1966) provided more. Further, if the area is made even smaller, the functions relating threshold to area and to duration are found to be similar (e.g., Björklund & Magnussen, 1979; Du Buf, 1987).

that the threshold versus area curve became steeper the less the background luminance, implying that in a graph like Fig. 2.2 the t.v.i. curve for a smaller patch would be shifted up and to the right, relative to the threshold curve shown in Fig. 2.2, in the same way as the isobrightness contours are. Chen et al. (1987) found that this was exactly true: t.v.i. curves for a small flash were the same as isobrightness curves for a larger one.[15] Together with the independence of brightness and area, this implies that the effect of repeating the whole family of increment measurements in Fig. 2.2 with small flashes would simply be to remove some of the lower curves, the remaining ones being unchanged except that the lowest would be relabeled "threshold."

2.3.7. *The Effect of Scattered Light.* A proportion of the light from any stimulus is scattered in the ocular media. The glaring stimuli represented in the upper left part of Fig. 2.2 probably scatter enough light into their immediate surround to outweigh the effect of weak steady backgrounds. It is difficult to estimate the effective proportion scattered because, although measurements of the physical scattering exist, we do not know what region of the skirt of scattered light around the test patch affects its contrast brightness. It may be quite a narrow region, but on the other hand a flashed background, which is what the scattered light provides, is more effective than a steady one. For decrements I cite evidence later that an effective 3.5% of surround light is scattered into the patch. I have assumed a smaller percentage for increments (2%), because the relative area of the source of the scattered light to that of the region over which it is effective is probably less, but the validity of the following arguments does not depend on any particular figure.

Such scattered light sets an upper limit to the retinal image contrast. The limit is shown in Fig. 2.7 in which the coordinates are retinal illuminances taking scattered light into account. A proportion k of ΔL is assumed to be added to L_b, so that the quantity plotted as abscissa in Fig. 2.7 is $L'_b = L_b + k\, \Delta L$.[16] Note that it is only ΔL, not absolute luminance, that has to be considered, because any uniform component common to both patch and surround is unchanged by scattering and therefore irrelevant. The effect of this transformation is that all the points in Fig. 2.2 that lay to the left of the contrast limits have been collapsed onto them in Fig. 2.7. Points more than a half a log unit to their right are little affected.

---

[15]This result was for cones. Sharpe, Whittle, and Nordby (1993) found a contrary result for rods.

[16]Fig. 2.7 uses L_{min} as abscissa for reasons given later in connection with decrements. Continue for now attending only to the increment data, for which L_{min} is the same as L_b. The algebra in this section applies to decrements if L_{min} is substituted for L_b.

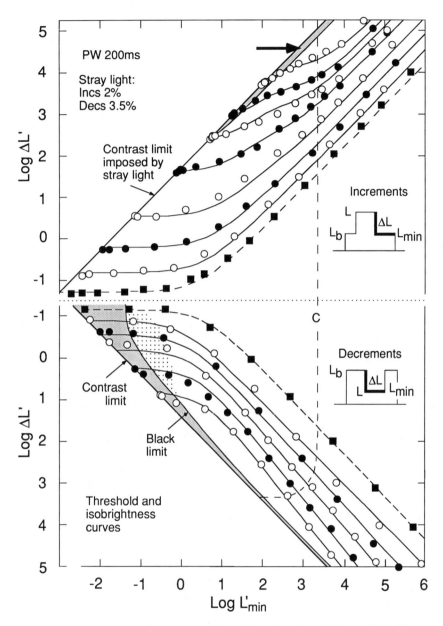

FIG. 2.7. Figure 2.2 (and Fig. 2.8) transformed to take account of stray light. The quantities plotted are the retinal illuminances that would result if k% of stray light were scattered from patch to surround, for increments, and from surround into patch, for decrements. The ordinate is $\Delta L' = (1-k)\Delta L$, and the abscissa $L'_{min} = L_{min} + k\Delta L$. L_{min} is the same as L_b for increments, but $= L$ for decrements. k is taken to be 2% for increments and 3.5% for decrements.

2. THE PSYCHOPHYSICS OF CONTRAST BRIGHTNESS 55

The striking difference between Fig. 2.2 and 2.7 shows that we must certainly pay attention to possible effects of scattered light. Might it be relevant to the rightward shift of the upper isobrightness curves? The answer seems to be that, although scattered light does not cause the shift, it does suggest a reason why such a shift is perceptually advantageous.

We have so far analyzed the shift into two components: the emergence of the upper branch corresponding to Stiles' π' mechanisms, and the shift to the right of each branch separately. However, neither component is affected by stray light: The upper branches are too far to the right, and the lateral position of the lower branches is unchanged even though they are severely truncated. We can see this by considering what effect the scattered light transformation has on a curve of the standard t.v.i. form, $\Delta L = c(L_b + L_0)$ (Equation 1 with $n = 1$).

We have $L'_b = L_b + k\Delta L = L_b + kc(L_b + L_0)$ and $\Delta L' = (1 - k) \Delta L$. Therefore $\Delta L'/(L'_b + L_0) = (1 - k) \Delta L/(L_b + kc (L_b + L_0) + L_0) = c(1 - k)/(1 + kc)$

That is, the effect is to decrease the Weber fraction by a factor $(1 - k)/(1 + kc)$, which shifts the curve vertically down but leaves its lateral position unchanged. The plateau is truncated at the contrast limit because L'_{min} cannot be less than kcL_0.

However, although scattered light does not cause the shift to the right, it does point to a perceptual advantage for it. If the isobrightness curves did not shift rightward, the contrast limit imposed by scattered light would make all lights brighter than a certain quite modest level indistinguishable. If the effective percentage scattered were 2%, the limit would be the brightness produced by about 2.2 log trolands in the dark. This is because in the ideal von Kries scheme the isobrightness curve corresponding to that level would have a Weber fraction at the contrast limit, and all greater brightnesses would be above the limit and so indistinguishable. In Fig. 2.7 they would all be collapsed onto the diagonal line representing the contrast limit. The thin stippled strip marked with an arrow shows the region to which the three highest isobrightness curves would all have been confined if in Fig. 2.2 they had obeyed the von Kries scheme (keeping the same ΔL_0 values). In fact, by virtue of the rightward shift, they remain quite well spaced out in Fig. 2.7, confirming that we can discriminate such high brightnesses. The stray light argument suggests that we can do so only because the decreased field sensitivity of the modified high intensity conditions of π_4 and π_5 reduces the masking effect of the halo of scattered light (see section 6.2).

Note that scattered light has a large effect on only those three highest sets of data points, whose lower branches look quite different in Fig. 2.7 because they have been so much moved down and truncated. Such brightness levels can be produced only by bright light sources, not by reflecting objects in a normal scene. It would not be surprising if they constituted a somewhat special case.

2.3.8. Adaptation to the Test Lights.
Another obvious explanation for the rightward shift is the adapting effect of the flashes themselves. This would increase with flash duration, and for flashes several log units above threshold might well be expected to outweigh the effect of weak backgrounds, thus prolonging the initial plateau that represents the background's having no effect, and shifting the curve to the right. This is not a cumulative slow adaptive process, such as pigment bleaching, because lengthening the interstimulus interval does not reduce the rightward shift (Whittle & Challands, 1969). Rapid self-adaptation to the flash remains a possibility, as Hayhoe and Wenderoth (1991) recently suggested. This may also be the basis of the Broca–Sulzer effect. However, Chen et al. (1987) showed that under the most obvious assumptions self-adaptation implies a vertical, not a rightward, shift (the algebra is the same as for stray light). Their preferred interpretation was that adaptation to bright backgrounds simply increased exponential gain with respect to ΔL. They showed also that this would lead to the improved visibility of small objects under bright light that is actually found. These explanations in terms of self-adaptation or increased gain are not in conflict with the analysis in terms of π mechanisms. For instance, adaptation could change both the response function and the relation between cone mechanisms.

2.3.9. Conclusions on Increments.
Suprathreshold isobrightness contours plotted as $Log\Delta L * Log L_b$ are similar to t.v.i curves (upper family of curves in Fig. 2.2). The t.v.i. curve follows $\Delta L = c(L_b + L_0)$, and the suprathreshold ones are at first a similar shape but shift to the right at higher brightness levels. This corresponds to a higher (exponential) gain with respect to $Log\Delta L$ on intense backgrounds.

The shift to the right has two components. One is a change in shape at higher brightnesses that can be interpreted as due to the intrusion of a second branch as in Stiles' two-color threshold experiments. Measurements of the field spectral sensitivities imply that the lower branch is Stiles' π_4 and/or π_5 and the upper the modified high intensity forms π_4' and/or π_5'. The latter have slightly narrower spectral sensitivities, higher absolute threshold, considerably less sensitivity to background light (L_0 is higher), and higher exponential gain than the lower ones (and, to anticipate, do not appear in the decrement data).

The second component is a rightward shift of each branch separately. This was abolished when the flash duration was less than about 50ms, and the test light was seen predominantly by a single π-mechanism (rod or cone). Each branch then followed an ideal von Kries scheme of multiplicative attenuation of all inputs by the same factor, $1/(L_b + L_0)^n$. ($n = 1$ for the present cone data but < 1 for rods.) This suggests that the rightward shift that persists when more than one mechanism is involved may be due to increased cooperation between them on brighter backgrounds. One would expect this to be associated with a change of spectral sensitivity for the separate branches, but this has not been looked for under these conditions. Might the change to the modified high-intensity condition be just an accentuation of this process at a particular background intensity?

The higher exponential gain on bright backgrounds may have two perceptual advantages. First, it may mediate our improved spatial resolution at high luminances (Chen et al., 1987). Second, it helps us to discriminate very bright lights by reducing the masking effect of the haloes of scattered light that they produce.

The basic measurements were made with a flash duration of 200ms. Lengthening duration beyond this, up to using steady lights, or changing the size of the flashes, had little effect on the pattern of results. Therefore they may be regarded as measurements of simultaneous contrast as well as of the adapting effect of a background.

2.4. The Darkness of Decrements

2.4.1. Decrements Compared with Increments.

The large experimental literature on vision tells us rather little about the similarities and differences of responses to increments and decrements. The different subcultures of psychophysics either concentrate on increments or ignore the distinction. The physiology of on and off responses looks relevant, but there is no consensus on how it maps onto the increment–decrement distinction. It is an area of surprising ignorance. Whittle and Challands (1969) published data for only increments. We did work with decrements, but for various reasons that work is only now being prepared for publication. I summarize some of it here. It turns out that there are both symmetries and important differences.

The very words we have to use confront us with this. The symmetry of *increment* and *decrement* suggests *brightness* and *darkness* as appropriate terms for the corresponding appearances. Darkness fits well into my conceptual framework. It is an acceptable opposite to brightness. Like brightness it is minimum at zero contrast, and its use reminds us that we are speaking of decrements. However, paradoxically, *lightness* is just as strong a candidate. The English language divides the continuum of grays into two so that *light* is the natural word—what linguists call the *unmarked* adjective—for the light half, and *dark* for the dark half. "Equating lightness" seems the better description of the matching task over most of the range, because all these decrements, except for the darkest, looked like surface colors, pieces of gray paper. Nevertheless, I use *darkness* because it makes my argument clearer but remember the claims of *lightness* and return to it in the next chapter.[17]

The quantitative data in Fig. 2.2 also show at a glance both symmetry and difference. The decrement threshold curve can be reflected and superimposed almost exactly on the increment one. But the suprathreshold curves show obvious differences. The first is trivial. Because you cannot take away more light than is present in the background, the data are all constrained to lie above the diagonal $\Delta L = L_b$ (dotted in Fig. 2.2). The second difference is more important: The higher contrast curves are very close-packed against the diagonal, even though the subjective intervals between the standard grays were roughly equal, as for the increments. This close packing represents a rapid rate of change of darkness with respect to LogΔL at high contrasts. Correspondingly, we see later that very small changes in LogΔL can be discriminated there. The discrimination data suggest that the quantity being discriminated in that region is L rather than ΔL, and in accordance with this we get a clearer representation of the decrement matching data if we use Log$L$ as one of the axes. This is done in Fig. 2.8 where the abscissa is L_{min}, which is L for decrements but L_b for increments. (We change the abscissa, not the ordinate, because ΔL must still be represented.)

This graph is more symmetrical than Fig. 2.2. This is the first indication that the expression $W = \Delta L / L_{min}$ may be useful for giving a comprehensive description of increments and decrements. We meet it in other contexts later.

The "black limit" in Fig. 2.8 is the threshold for detecting light in a "black" decrement of 100% contrast (section 5.2 and Fig. 2.14). Therefore, in the heavily stippled region the light in the decrement cannot be seen. Changing L there has no effect. Changing ΔL, which is what was done to try to make a match, changes the brightness of the surround, a halo of which can be seen around the patch even in the HSD, but the patch remains more or less black, just becoming clearer when ΔL is larger.[18] When L_b is small, so that the decrement is taken away from a background that is itself scarcely visible, the patch looks like a black object seen through fog. As ΔL (and therefore L_b) is increased, the fog clears, but it is hard to take a

---

[17]*Dark* is a strong quality like *bright*. They are the appropriate opposites in a bipolar or opponent-colors framework. *Subdued* or *dim* are the natural opposites to *bright* when describing colors or light sources, in a monopolar framework. There is no settling this. Different frameworks are appropriate to different aspects of vision.

[18]Various methods were used in the decrement matching experiments, to check that unexpected features of the results, such as the difficulty in matching under some conditions or the slopes being > 1.0, were not specific to particular methods. Sometimes L was held constant, sometimes L_b. Method of Adjustment, Constant Method, or Interleaved Staircases were used

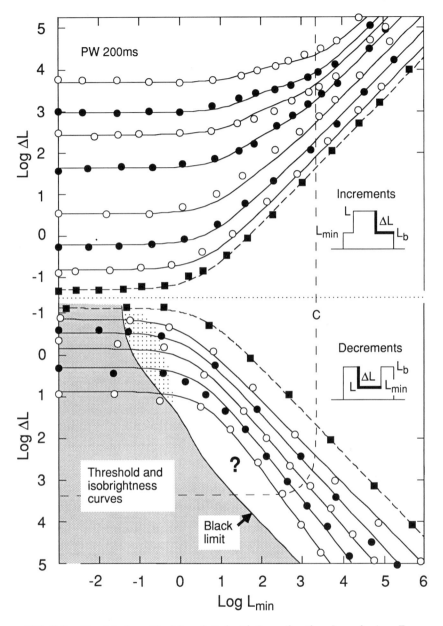

FIG. 2.8. The data from Fig. 2.2 replotted with L_{min} rather than L_b as abscissa. For increments $L_{min} = L_b$ so their representation is unchanged. For decrements $L_{min} = L$, so their abscissa is now the luminance of the patch rather than of the background. Here and in Fig. 2.7 the lines through the points follow the single-parameter form of Equations 2 and 3. The dashed line C is line C of Fig. 2.2 transformed to the new abscissa, to show the shape of a constant L_b line in the new coordinates. As L_b changes, this line slides up and down the negative diagonal $\Delta L = L_{min}$ of the lower half. The "black limit" is the threshold for detecting L in a black decrement (Whittle, 1986). The lightly stippled region in this and other figures marks a region where matching is difficult.

subjective attitude and see the fog as grayness belonging to the patch, which is what is necessary to match it to a standard that is a gray surface color.[19] Results have been variable even when subjects have tried to maintain the required attitude, and the mere fact that such an effort is necessary, which is not at all the case anywhere else in either the decrement or increment domains, shows that something special is happening here. To see these phenomena convincingly, a haploscopic display is necessary, to allow one eye to be adapted to a low luminance and the other to a luminance high enough to see good grays. Although a black limit was inevitable—there must be a threshold for seeing the light in a black decrement—these problems persisted into the lightly stippled region to the right of the black limit.

This therefore constitutes a third difference between increments and decrements. Whereas for increments matching was satisfactory over the whole $\Delta L^* L_b$ space, for decrements there are no-go regions. Only if the stippled regions are avoided is the condition for the validity of the whole enterprise reliably true: that it is the darkness of the opposite-eye standard that determines the curve. I have verified this for standards that lay outside the stippled regions, using a transitivity test as described in section 2.3.1 for increments. For standards within those regions, however, it was true only if the subject maintained the difficult subjective attitude of seeing the "fog" as grayness belonging to the patch. Because matching is so difficult there, little weight can be attached to the precise position of the

---

as psychophysical techniques. All these methods gave essentially the same results. The black limit was measured by a spatial 2AFC task.

The physical makeup of the stimuli was important in generating results like those in Fig. 2.2. To be able to work close to threshold, an essential condition was to control ΔL sufficiently accurately to match the finely graded visual response to it. For that, in the Maxwellian view equipment used, the stimuli had to be physically made up as an increment ΔL added to a uniform background. It was almost impossible to do it by presenting L in a hole in L_b, as work on brightness contrast has traditionally done, because all kinds of imperfections in the apparatus—residual edges, nonuniform luminances, and so on—become obtrusive in the zone where L and L_b are nearly equal. For decrements, the stimuli had to be made up as a surround annulus of luminance ΔL superimposed on a uniform background L. This was serendipitous, because it gave independent control of ΔL and L and so allowed the discovery of, among other things, their relative importance in different parts of the decrement range. However, it also raised the doubt that these results might be an artifact of the way the stimuli were controlled. This doubt can easily be laid to rest nowadays by generating the stimuli on a computer monitor, on which it is L and L_b that are manipulated. The results, where comparable, agree with the earlier measurements.

[19]This is a perceptual ambiguity that the haploscopic display does not remove, because it depends on the luminances in one eye rather than the relationship between the surrounds. It is probably a factor in the controversy over whether blacks become blacker as the overall illumination is increased. The ambiguous appearance allows subtle features of the experimental instructions or context to influence subjects' judgments, which could well generate the interstudy disagreements noted by Jacobsen and Gilchrist (1988a).

data points. Those given in Fig. 2.2 probably average different strategies, such as maintaining the subjective attitude or matching salience. In terms of Equation 1, the parameter L_0 becomes uncertain because it depends on the position of the plateau, which is mostly within the doubtful region.

Where the matches can be made satisfactorily, the isodarkness curves in Fig. 2.8 are straight along most of their length. This is a fourth difference from increments: The higher curves show neither the two-branched shape nor the shift to the right. If my interpretation of the upper branch is correct, then its absence here suggests that Stiles' modified high-intensity condition does not respond to decrements. The longer linear portion of the decrement curves would accord with the generalization sometimes made that decrements obey the ratio rule more closely, except that the slope is not exactly 1.0. It is 1.0 at threshold, but increases with the distance from threshold. This is clear in Fig. 2.8, just visible in Fig. 2.2, and a large effect in terms of Michelson contrast, as we see later (Fig. 2.13). It is a fifth difference from increments, for which I have never found slopes of more than 1.0. We see later, however, that this difference disappears when we look at the data in terms of L rather than ΔL (section 4).

Note that any interocular filling-in of the standard decrement by the brighter backgrounds haploscopically superimposed on it in the HSD would have the opposite effect on the slope. The standard would become lighter the greater the value of L_b in the variable eye, so the matching ΔL in that eye would decrease with L_b, not increase.

2.4.2. One Equation for the Decrement Data.

Individual isodarkness curves are all well fit by a generalization of Equation 1:

$$\frac{\Delta L}{(L_{min}+L_0)^n} = c \qquad (3)$$

The only change from Equation 1 is the substitution of L_{min} for L_b. To generate a single equation to fit the set of isodarkness curves, ΔL_0, the value of ΔL when $L_{min} = 0$, which was used for increments, is not such a suitable parameter because the plateau height is unreliable. I used instead ΔL_3, the value of ΔL when $\text{Log}L_{min} = \text{Log}L = 3.0$. A single-parameter form of Equation 3, $\Delta L = g(L_{min}, \Delta L_3)$, was then generated as described for increments in section 2.3.3. The continuous lines through the decrement data in Fig. 2.2, 2.7, and 2.8 were drawn from this equation, with the appropriate transformation in the case of Fig. 2.2 because the abscissa is then $\text{Log}L_b$.[20]

---

[20]The equation was $\Delta L = 0.016 * \Delta L_3^{0.449}(L + 0.22 * \Delta L_3^{0.31})^{\text{Log}(3.97\Delta L_3{}^{\wedge}0.184)}$. This does not include the threshold curve, which follows Equation 1. It can be included if $\text{Log } L_0$ is allowed to be a quadratic function of $\text{Log}\Delta L_3$, because L_0 increases above threshold and then decreases. This would be a theoretically important point, because I want to claim that the threshold curve

2.4.3. The Effect of Scattered Light. The perception of high-contrast decrements must be affected by light scattered in the ocular media (e.g., Whittle, 1986). This was shown in the lower half of Fig. 2.7, which was based on the assumption that 3.5% of the surround luminance is scattered into a decrement patch. This figure was derived by comparing thresholds for a small probe flash in a black decrement and on the neighboring surround (Whittle, 1986), for a range of values of L_b. The effect is to impose a contrast limit, as for increments, and to narrow the blank region between the black limit and the nearest isodarkness curve. Because light is scattered on a purely proportional basis, the slopes of the curves are little changed.

2.4.4. Duration and Darkness. Matching decrement flashes of unequal duration generates curves like those shown for increments in Fig. 2.5. Suprathreshold there is a modest Broca-Sulzer effect, as found by Björklund and Magnussen (1979). The effect on isodarkness curves is shown in Fig. 2.9. These were derived from an experiment in which L was constant during each run, and the subject adjusted ΔL to set threshold or make lightness matches at nine flash durations between 10ms and 3s. Conditions were otherwise like those of the increment experiments for variable durations (Fig. 2.5). In Fig. 2.9, the axes refer to the 500ms data. Pairs of curves for other durations were moved to make all threshold curves coincide with that for 500ms. Therefore the graph shows how the position and shape of an isodarkness curve changes relative to threshold as duration is increased. The interesting point is that the changes are opposite to those seen with increments. With increments, as duration is increased the suprathreshold curves move to the right and become less steep. With decrements, they move to the left, if anything, and become steeper.

The explanation presumably lies in the different dynamics of rapid light and dark adaptation during the flash. I suggested that the plateau in the increment curves at low background luminances may lengthen as ΔL or t is increased because of the increased self-adapting effect of bright flashes. The opposite, if anything, might be expected for decrements: that they become more sensitive to the background luminance when they are of higher contrast or longer, because of the greater dark adaptation that then occurs within the test patch.

Differences between increment and decrement data are summarized later (section 6.8) after the results of other experimental techniques have been discussed.

---

belongs to the isodarkness family, were it not for the fact that the value of L_0 depends on the dubious matches in the stippled regions. The reliable parts of the isodarkness curves, including the whole threshold curve, can be well fit by a version of Equation 3 with constant L_0, as in the ideal von Kries scheme. The exponent n, however, has to be allowed to vary.

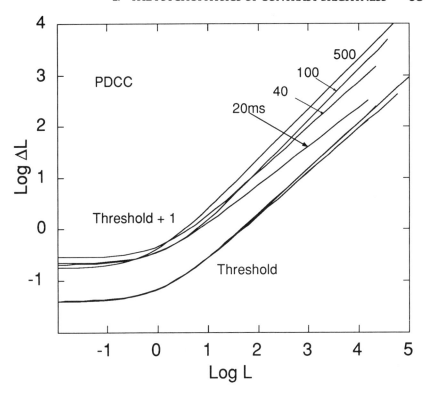

FIG. 2.9. Threshold and isodarkness contours at different flash durations, under steady fixation. Unpublished data, for one subject; data points are omitted, for clarity. All "Threshold + 1" curves were fit to matches to the same 200ms left-eye standard, which was 1.0 log unit above its detection threshold. The scales apply only to the 500ms curves.

2.5. Matching Apparent Contrast

The difficulty of matching at low luminances raises a serious problem for a Stilesian analysis of decrements in terms of t.v.i. curves (Fig. 2.3a) because it makes the plateau height uncertain. One way to avoid this is to match apparent contrast rather than brightness. If a standard with contrast of opposite sign is used, so that decrements are matched to increments and vice versa, there is no possibility of unwittingly making a brightness match. For low and medium contrasts such matching is easy (Burkhardt, Gottesman, Kersten, & Legge, 1984; du Buf, 1987). The judgment can be made quickly and directly, not by means of consciously judging the separate brightnesses of patch and surround. At high contrasts a ceiling is reached for decrements—they cannot be made blacker than black, but increments can be made as bright as one can bear—so when matching apparent contrast one cannot explore as much of the (L, L_b) space as when matching brightness. Preliminary results with this method are similar in form to Fig. 2.8 but closer to the ideal von Kries scheme, with $n = 1$ for both increments and decrements. This suggests that some of the results discussed previously may be specific to judging brightness or color rather than general features of retinal contrast coding.

Although the plateaux of iso-apparent-contrast curves may be better defined, the differences from the brightness matching data prevent us from substituting these plateaux for the ill-defined isodarkness ones. The latter therefore remain uncertain, and little weight will be put on them in the following analyses.

2.6. Flashing the Surrounds Instead of the Patches

Another variant of the basic technique is to flash the surrounds rather than the test patches. An incremental surround flash leaves a decrement (spatially defined) in its center, and a decremental surround leaves a spatial increment. The brightness of the center patches so produced varies with ΔL and L_{min} just as when it is the center patch that is flashed. One takes this kind of behavior for granted when working with these phenomena, but it does, of course, strongly reinforce the idea that all these appearances are a function of *relative* luminance. If such stimuli are used for brightness matching, the form of the isobrightness curves is predictable from the appearance and spatial contrast of the test patch rather than from whether the surround was incremented or decremented;[21] that is, darkness matches made by varying the luminance of an incremental surround flash produce isodarkness curves like the decrement curves in Fig. 2.8, and correspondingly for brightness matches, mutatis mutandis. This is not surprising because steady lights are the limiting case of both techniques, and we already know that they produce isobrightness contours like those when the test patch is flashed. However, the fact that decrement data can be produced by an incremental surround flash and vice versa does show that increment–decrement differences are not due simply to different responses to the onset and offset of light per se.

3. SCALING CONTRAST BRIGHTNESS: THE CRISPENING EFFECT

Brightness matching, the basis for all the data so far discussed, sidesteps the thorny problem of "measuring brightness." But even in describing the matching data, I talked about the "gain of brightness with respect to ΔL" and "the subjective intervals between the grays." We want to say such

---

[21]These experiments were carried out in my laboratory by Roger Springbett.

things, and preferably with quantitative precision. We cannot regard them as only shorthand references to hypothetical internal signals, as a physiologist might be tempted to, because the internal signal model would be evaluated partly with respect to such statements. We have to be able to evaluate them in their own terms. Now of course there is a large and controversial literature that claims to measure brightness. It seems to be of practical value but has made little contribution to our knowledge of visual mechanisms.[22] The defects of one popular method, magnitude estimation, trenchantly criticized by Poulton (1989), are a major reason for this.

I have used instead a method that seemed to me to make fewer assumptions and certainly yields data with far less variability (Whittle, 1992). Subjects set an equal-interval scale, with all the stimuli visible at once. They saw 25 circles on a computer screen, in a gray surround, with the end circles fixed at black and white.[23] Their task was to set the luminances of the other circles to make all the brightness intervals look the same. This amounted to setting the 24 luminance steps to be something like "equally noticeable at a glance." Individual subjects set smooth scales and the variability compared favorably with much more time-consuming discrimination threshold measurements. Munsell, Sloan, and Godlove (1933) had used the same method and liked it for the same reasons. Poulton (1989, ch. 4) recommended equal-interval scales as the baseline to evaluate other scales.

The scale set is shown in Fig. 2.10. The steepening at the background luminance is what Takasaki (1966) named the *Crispening Effect*. It is neglected in many studies of brightness scaling, although described by von Bezold (1874) and well discussed by Wyszecki and Stiles (1982). Semmelroth (1970) described it with an equation combining power laws in L and ΔL, writing that "the idea is that there is a direct sensory response to stimulus differences . . . This is in accord with the effects of simultaneous brightness contrast" (p. 1686). The steepness at $L = 0$ (filled squares) is the same phenomenon as the close packing of the isodarkness curves against the diagonal in Fig. 2.2.

The close relation between the shape of this scale and the matching data is most concisely shown by the fact that the scale is well described by the equation

$$\text{Brightness} = a\text{Log}(1 + bW') \qquad (4)$$

---

[22]The recent review of visual psychophysics and neurophysiology edited by Spillmann and Werner (1990), by some 70 authors well known in the field, contains, as far as I can see, no references at all to scaling or magnitude estimation.

[23]The circles subtended 2° at the viewing distance of 60cm, with intervals of about 1° between them. They were seen on a screen 19° times 14°, arranged in a rough spiral to avoid jumps at the end of lines. The end circles had the maximum and minimum luminances that the monitor would yield (74.1 and 0.27 cd/m^2).

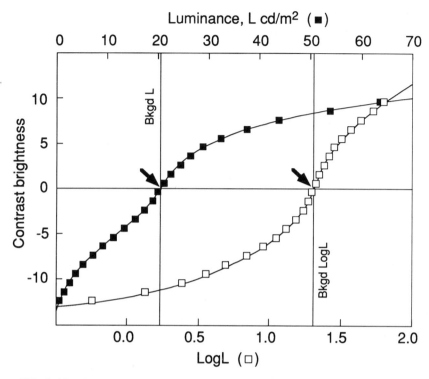

FIG. 2.10. An equal-interval brightness scale for 25 gray circles on a gray background, plotted against both linear and log abscissae. The ordinate unit is just the ordinal number of the circles with the origin shifted to make brightness zero at zero contrast. The inflections marked by arrows are the Crispening Effect. Average of five subjects, from Whittle (1992). Units in this graph are cd/m^2.

where $W' = \Delta L/(L_{min} + L_0)$. The line through the points in Fig. 2.10 follows Equation 4. This suggests that brightness should be constant if W' is constant, which we have already found to be true because it is to a first approximation the shape of the isobrightness contours in Fig. 2.2 and 2.8.[24] It is also an encouraging pointer toward the possibility of a quantitative description of contrast brightness encompassing both matching and scaling data. The shape of the brightness scale is discussed further in sections 5.2 and 6.3.

---

[24] $W (= \Delta L/L_{min})$ has to be generalized to a form like W' if it is to encompass the case where $L_{min} = 0$. The values of the coefficients were: $a = -7.07$ for decrements and $+8.22$ for increments and $b = 6.58$ for both (Whittle, 1992). Both vary somewhat with L_b. The coefficient b, which determines the slope of the scale at $L = L_b$ (the steepness of the Crispening Effect), is probably very dependent on the constraints under which the scale is set (endpoints, number of steps). It is for these reasons that Equation 4 only *suggests* that brightness should be constant if W' is constant.

To describe this scaling procedure as measuring brightness would make too general a claim and raise too many conceptual problems. But it does seem a promising method for giving operational meaning to statements about the rate of change of brightness and the relative size of brightness intervals.

4. MAPPING CONTRAST BRIGHTNESS SPACE

We now have a basic tool kit for mapping the three-dimensional space defined by the quantities Brightness, L and L_b. We have equations for isobrightness contours and we can use Equation 4 to assign numerical brightness values to the contours. For many reasons these maps can be only tentative, most obviously because they depend on extrapolating from one brightness scale set over a modest luminance range under specific conditions. However, later we see that the form of the scale agreed well with discrimination data obtained under very different conditions, and the resulting maps are similar to those obtained by other investigators in other ways. They probably have some generality.

At the very least, this space provides a useful representation. In the context of this chapter, it helps us to see convergences between the results of the different experimental methods. Indeed, one could say that the central project of the work was to explore this space by different methods. That, however, would beg the question of whether such a unified quantitative description is possible.

Figure 2.11 shows a contour map. The lines are isobrightness contours following the one-parameter forms of Equations 2 and 3, which I refer to collectively as the "isobrightness equations." The coefficients used were those that gave the best fit in Fig. 2.2 and 2.8, so their shape accurately reflects the matching data. The numbering and spacing of the contours, on the other hand, is based on the equal-interval scale described in the previous section. The numbers are the value of B/B_0, where $B = a\mathrm{Log}(1 + bW')$ (Equation 4), and B_0 is the value of B at increment threshold; that is, the numbers are brightness on the equal-interval scale, using the value of brightness at increment detection threshold as the unit. This unit is approximately half the coarse JNDs set in the equal-interval scale, and twice the fine ones measured with the forced-choice technique described in section 5. The luminance range and value of L_b over which that scale was set is shown by the thick vertical line. B and B_0 were evaluated along that line. It can also serve as a scale object for the map: Its top is nearly the brightest white and its bottom nearly the darkest black that could be produced in a mid-gray surround on a computer monitor. The dashed rectangle encloses the most commonly used photopic lumi-

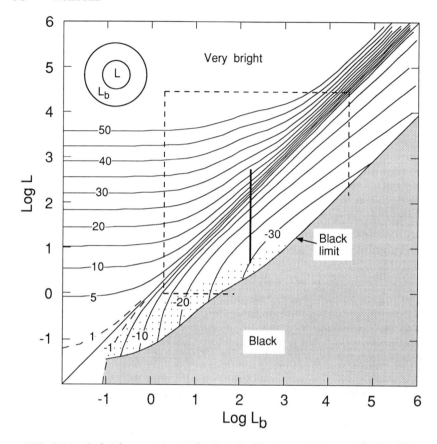

FIG. 2.11. Isobrightness contours for a patch of luminance L in a surround L_b. The brightness values were determined by an equal-subjective-interval scale set along the range marked by the thick vertical line. Increment threshold was assigned value 1. The dashed rectangle encloses the most commonly used photopic luminance range. Units are trolands.

nance range. This is taken from Le Grand's figures for "light colored objects" from "feeble interior lighting" to "daylight out-of-doors" and extended downwards by a log unit to include dark objects (Le Grand, 1957, p. 83). The stippling has the same meanings as in Fig. 2.8.

Several features of this map are striking. The approximate symmetry about the diagonal and the 45° slope of many of the contours show the importance of luminance ratios. The high density of contours around the diagonal is the Crispening Effect, where brightness changes very rapidly with respect to both Log$L$ and LogL_b. This shows the importance of ΔL. The two features together show the dependence on contrast. Although the first impression is of approximately 45° slopes, notice that actually both increment and decrement contours have slopes somewhat less than 45° over most of the range enclosed by the dashed rectangle; that is, Wallach's ratio rule does *not* hold accurately for contrast brightness at normal illumination levels.

This result is familiar in the literature in two different forms, depending on whether the discussion is of brightness contrast or lightness constancy. In the former, it is commonly reported that the induced brightness change produced by varying L_b is less than the direct change produced by the same variation in L. This was clear, for example, for all modulation rates in the data of Magnussen and Glad (1975b). In the demonstration described in section 1.1 the brightness modulation produced by changing L is greater than that produced by changing L_b, unless the contrast range is small enough to lie within the region near the diagonal in Fig. 2.11 where the slopes are exactly 45°. In the context of constancy, it is commonly reported that lightness constancy as illumination is changed is less than complete, like most constancies.[25] This corresponds to the slopes being less than 45° because that implies that increasing L and L_b in proportion (moving at 45°), which is what happens in a natural scene of reflecting objects when overall illumination is increased, will take you to higher contours. This is our common experience: When illumination is increased, everything gets a bit brighter.

Drawing such maps with different axes shows up different features of the data. How much increment–decrement symmetry? Are there horizontal lines showing that brightness is invariant with the ordinate? Are there bunched-up contours as around the diagonal in Fig. 2.11? If so, the axes are probably failing to represent some functionally important quantities (ΔL, in the case of Fig. 2.11).

Figure 2.12 represents the same quantities as Fig. 2.8, ΔL and L_{min}, except that they are now combined in the ordinate $W = \Delta L / L_{min}$, so that the threshold curves are parallel and approximately horizontal at medium to high luminances. In some ways this is the best contour map of brightness. Because $\log W$ is the ordinate there is no bunching, and the increment and decrement curves are fairly symmetrical. W incorporates ΔL and both L_b, for increments, and L for decrements: all the relevant quantities. The impression that the "ratio rule" is followed precisely over a large part of the map, where the contours are horizontal, is somewhat misleading, as we have just seen, because this central region all represents very low-contrast stimuli. This map does, however, make it clearer than the previous one that the range over which decrements obey an approximate ratio rule is greater than that for increments.

---

[25]This is taking us away from the special situations that produce contrast brightness, but we see in the next chapter that constancy experiments in which lightness matches are made between displays differing only in illumination do in fact produce results that agree with the isobrightness contours.

Figure 2.13 has Michelson contrast as ordinate. The contours are now bunched at the top because Michelson contrast is constrained to be ≤1 and so obscures our good discrimination of high luminances. This map emphasizes the observation that for constant contrast decrements become much lighter at high values of L_b. At a background luminance of about 1.5 Log td brightness and darkness are approximately symmetrical (equal for equal M), up to contrasts of about 0.6. At higher luminances this is not at all the case. Such symmetry was reported for subjective contrast by Burkhardt et al. (1984) and du Buf (1987) and for discrimination threshold by Whittle (1986).

In a three-dimensional space, maps can be made in three orthogonal planes. Those presented so far have all been versions of the L, L_b plane. Before looking at other planes, we should consider a further type of measurement.

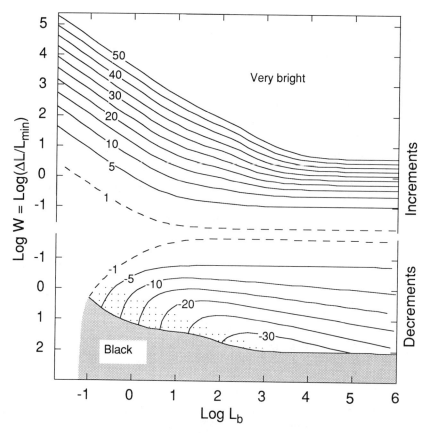

FIG. 2.12. Isobrightness contours for axes Log W ($\Delta L/L_{min}$) versus LogL_b.

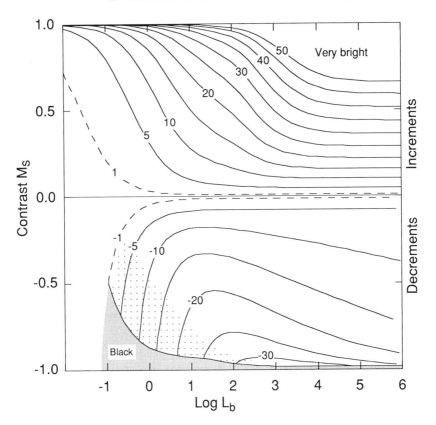

FIG. 2.13. Isobrightness contours for axes Michelson contrast versus Log L_b.

5. DISCRIMINATING CONTRAST BRIGHTNESS

5.1. Measuring Discrimination Threshold

Brightness scales provide one approach to estimating the rate of change of brightness in the L, L_b space. Another is to measure discriminability, as has been recognized at least since Fechner's attempt to construct a scale of sensation by integrating just noticeable differences (JNDs). Accordingly, I measured luminance discrimination between spatially separate lights such as were used for matching and scaling (Whittle, 1986; Whittle & Swanston, 1974). The results confirm by this very different route several features of both the matching and scaling data.

To yield this outcome, the luminance discrimination task has to be carefully chosen. It is crucial that the patches are spatially separate. Discrimination of contiguous patches (or the detection of an increment within one of them) depends on an edge-detection process that has little to do with the brightness of the two regions, because that is mainly determined by their outer edges (Cornsweet & Teller, 1965). To try to relate the discrimination of contiguous patches to brightness would be like trying to study the apparent length of lines by measuring the discriminability of two parallel lines aligned at one end. The discrimination would be based entirely on the misalignment of the other ends, ignoring the rest of the line, so the threshold could tell us nothing about apparent length. Failure to appreciate this has vitiated some previous attempts to link discriminability and appearance (see the review in Le Grand, 1957). Nachmias and Steinman (1965) made this point clearly in the context of brightness.[26]

The principle is that for discrimination measurements to be relevant to judgments of appearance both tasks must be based upon the same subjective dimension of the stimulus and hence on the same neural events. Of course, one can just tell subjects to discriminate on the basis of a particular dimension, such as brightness, but not only is this a difficult instruction to obey when other cues are available, it also removes most of the point of measuring discrimination thresholds, which is to obtain confirming evidence from a rigorous performance measure. The constraint has to be imposed, not by command but by careful choice of stimuli. In the present experiments contrast brightness was the most obvious cue over most of the range. At or below detection threshold simple visibility or salience was used. Even though brightness was the dominant cue, the measurements can be thought of as either luminance or contrast discrimination because luminance and contrast were covarying. Contrast discrimination of gratings yield similar data at low and medium contrasts (see Whittle, 1986). This question is taken up again in section 10.1.

Discrimination of spatially separate lights is quite different from the usual increment threshold because it is discrimination between two transients rather than detection of one. The importance of this difference is confirmed by the characteristically different t.v.i. functions produced. Detection yields a monotonic increasing function like that in Figure 2.3(a), with a gradual transition to Weber's Law, whereas discrimination yields the dipper function (Fig. 2.14b) in which threshold has a minimum around detection threshold but obeys Weber's Law thereafter, without the gradual transition. This is true in all sensory modalities (Laming, 1986; Whittle & Swanston, 1974).

---

[26]In spite of this, the relationship between brightness and JNDs is sufficiently robust that approximate agreement can be shown even when the JND is not measured in what now seems the most theoretically correct way. We see this for instance in Heinemann (1961). My brightness matching and discrimination experiments can be seen as a replication of Heinemann's with a few changes that generate cleaner results in some respects and show up some new relationships. The explanation for the robustness of the brightness–JND relationship may lie simply in the pervasiveness of Weber behavior.

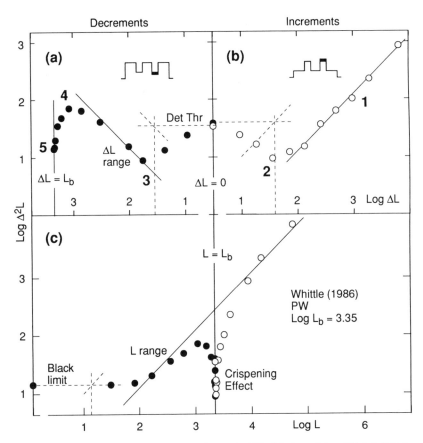

FIG. 2.14. Discrimination threshold ($\Delta^2 L$: marked black in insets) from Whittle (1986) for 200ms increment and decrement flashes like those used for the brightness matches of Fig. 2.2. (a) decrements and (b) increments with Log ΔL as abscissa (increasing from right to left in (a)). (c) shows the same data plotted with Log L as abscissa. The diagonal lines have slope 1.0. These measurements were made along the dashed line C in Fig. 2.2, and the numbers 1 to 5 in the two figures mark the same ΔL values.

5.2. How Discrimination Threshold Varies with ΔL and L_b

Figure 2.14(a) shows how discrimination threshold varies with LogΔL, with L_b constant. To understand these data it may be helpful to relate them to Fig. 2.2. Figure 2.14(a) can be thought of as a vertical cross-section of Fig. 2.2 along line C, showing at each point an equivalent to the error bars that are absent in Fig. 2.2.[27] The numbers 1 to 5 in the two figures mark

---

[27]The data of Fig. 2.2 were not collected in a way that gave meaningful error bars. Of course the discrimination threshold is not exactly equivalent to the variability of brightness matching,

corresponding ΔL values. The following description takes them in order from right to left: (1) Suprathreshold increments obey Weber's Law. This corresponds to the more or less even spacing of isobrightness contours in Fig. 2.2 and 2.12. (2) Around detection threshold increments follow the dipper function. The points between (2) and (3) fall in the triangular subthreshold area between the increment and decrement families of curves in Fig. 2.2. (3) Decrements also follow the dipper function and start along the same Weber line, but they reach a peak (4) where $\Delta L = L_b/2$, and as ΔL approaches its maximum value L_b (5) the threshold drops steeply. This improvement in discrimination corresponds to the close packing of the isodarkness curves along the diagonal in Fig. 2.2. In Fig. 2.14(c) the steep drop for dark stimuli is stretched out by using Log L as abscissa. The form of the curve suggests what is happening: Over a certain range of these dark stimuli (the "L range"), threshold is proportional to simple luminance, L, not to differential luminance, ΔL.[28] The same maneuver of plotting Log$L$ as abscissa in Fig. 2.8 spaced out the isodarkness curves. For luminances smaller than the L range, the curve in Fig. 2.14(c) flattens out, forming a plateau whose height is the "black limit" shown in Fig. 2.7 and 2.8.

The division of the decrements into ΔL and L ranges parallels the linguistic distinction between light and dark grays (section 2.4.1). Figure 2.15 may make the division more vivid. This is a cross-section of a black decrement with stray light, detection threshold, and three values of discrimination threshold all drawn to linear scale. The discrimination threshold is small at the two extremes—detection threshold and the "black limit"—and large at the midpoint. We are good at comparing almost full or almost empty buckets, but not so good when they are half full. It is plausible that the good discriminations are based on devices for detecting on the one hand the difference between L and L_b, and on the other the absolute amount of light, and that we do not do so well in the mid range because neither device is optimal there.

The very marked cusp at $L = L_b$ in Fig. 2.14c corresponds to the Crispening Effect, which was seen as the inflection in the brightness scale in Fig. 2.10 and as dense contours around the diagonal in Fig. 2.11. Plotting

---

but they are both indices of the rate of change of brightness. The threshold, $\Delta^2 L$ $(=\Delta(\Delta L))$, is the difference (marked black in the insets to Fig. 2.14a and 2.14b) between the luminances of the two patches that gives 71% correct judgments as to which is greater. The task was a spatial 2AFC, and subjects were told whether they were right or wrong after every trial. The size, duration, and position of the stimuli were approximately as in the matching experiments of Fig. 2.2, though in the discrimination measurements both patches were seen by the same eye. Measurements were made on all the four background luminances marked A to D in Fig. 2.2.

[28]This is described rather misleadingly in Whittle (1986) as showing that discrimination then depends on *absolute* luminance. The difference is between dependence on L and on ΔL. The effects of both are probably relative to adaptation level.

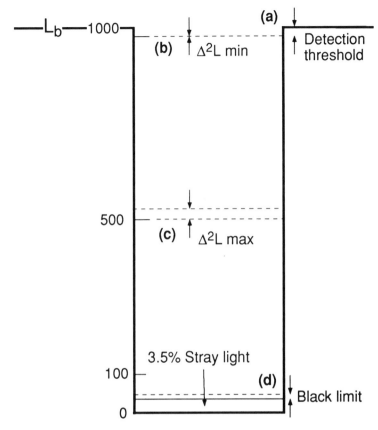

FIG. 2.15. A cross-section of a black decrement with $L_b = \Delta L = 1000$ td, and $L = 0$. Stray light (3.5% of ΔL; see section 2.4.3) and four threshold values are all drawn to linear scale. The thresholds, which can also be seen in Fig. 2.14(a) and 2.14(c), are: (a) decrement detection threshold ($\Delta L = 22.1$). (b) The minimum discrimination threshold, $\Delta^2 L$, which occurs near detection threshold ($\Delta^2 L = 3.5$, about the width of the thin lines in this diagram). (c) The maximum $\Delta^2 L = 29.0$, when $L = \Delta L = L_b/2 = 500$. (d) $\Delta^2 L = 9.9 =$ the black limit, the threshold for detecting light added to a decrement of 100% contrast. Of course, only (d) is actually measured in a decrement of the contrast shown; the other values are shown for comparison.

LogΔL in 14a and 14b stretched out this cusp into a W-shape: a double dipper. The fact that the outer sides of the W follow Weber's Law with respect to ΔL confirms Semmelroth's (1970) explanation for the Crispening Effect: that ΔL is the relevant variable here. The same result was clear in some of the scaling data (Whittle, 1992). The same conclusion—that it was luminance *differences* (ΔL) that were being discriminated—was also forcefully demonstrated by the effect of changing L_b. Even when this altered the total luminance L by orders of magnitude, the Weber fraction $\Delta^2 L/\Delta L$ was

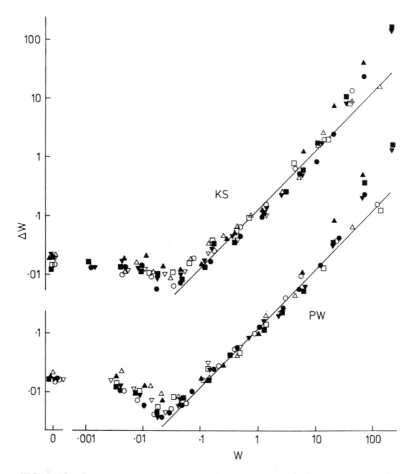

FIG. 2.16. Discrimination thresholds from Whittle (1986, Fig. 2.9), for two subjects plotted in terms of $W = \Delta L/L_{min}$. Hollow symbols represent increments, filled ones, decrements. The different symbol shapes correspond to different background luminances: upright triangles, 1.35 log td; squares, 2.35; circles, 3.35; inverted triangles, 4.35.

almost unaffected (Whittle, 1986). Clearly, the discrimination was based on ΔL, not on L. The background was completely discounted.[29]

The discrimination data can be economically described in terms of W. Figure 2.16 shows that the second-order Weber Law, $\Delta W/W$ = constant,

---

[29]The subthreshold side of the dipper is of course the *pedestal effect* of Leshowitz, Taub, and Raab (1968), which is often explained in terms of a positively accelerated response function to luminance (Cohn & Lasley, 1986). Here I am emphasizing the other side: the Weber Law with respect to ΔL. The existence of the pedestal effect follows logically from two facts: Weber's Law continues down to detection threshold, and the Weber fraction is less than 1.0. Which side of the dipper will provide the more fundamental explanation for the effect is an open question.

describes to a good approximation both increment and decrement data, and for the latter encompasses both ΔL and L ranges. Thus W provides the basis for simple descriptions of all three kinds of measurements: matching, scaling, and discrimination.

6. CONVERGENCES AND DIVERGENCES

6.1. How Much Convergence Is There Between the Three Methods?

The three methods of matching, scaling, and discriminating are convergent in two senses. First, they provide different but complementary perspectives on the same subject matter. Thus, for example, the matching and discrimination data illuminate each other, or the matching and scaling data can be combined to generate the contour maps. These complementary perspectives reveal the same features occurring in all three types of data. The most obvious are: (a) the increment–decrement separation; (b) an approximate Weber's Law: Brightness depends on contrast; (c) the ΔL range: the Crispening Effect; (d) the L range: the dependence of the appearance and discriminability of high-contrast decrements on L rather than ΔL. These are not all shown equally clearly by all three methods, but they are all discernible in each set of results to some extent. This is shown most concisely by the wide applicability of the ratio W.

The dependence on contrast is clearest for low and medium contrasts. In Fig. 2.11 contrasts less than 50% correspond to a band extending only half a log unit on each side of the diagonal. Within such a band the dominant pattern is of contours at about 45°. For higher contrasts there are two pieces of evidence, one for increments and one for decrements, showing a greater importance for luminance. For increments, the long plateaux of the upper isobrightness contours in Fig. 2.2 show that L_b has little effect until it is very intense. The fact that the plateaux disappear in Fig. 2.7 shows that the mechanism of this is largely the scattering of light in the eye. For decrements the L-range, which is clearest in the discrimination data, also shows the importance of luminance. Data for all contrast levels can be accommodated in one scheme by using appropriate contrast expressions as in Equations 2 and 3, but it would be misleading to say that this shows a dependence on contrast over the whole range.

The recurrence of the variable W is one basis for my claim in the introduction that "The agreement is close enough for the outlines to appear of a simple quantitative model to accommodate all the different kinds of data." If that could be achieved, it would be convergence in a second, stronger, sense. As it is, the same *form* of expression recurs, but more precise quantitative agreement showing that not only the form but also the value of the parameters was the same is elusive. We see some instances of it shortly (sections 6.2 and 6.3), but not as a general finding. Further work may reduce the discrepancies, but I do not think that this strong convergence should be thought of as the main criterion for judging the success of the enterprise. There are good reasons why each method should have its own peculiarities and therefore good reasons for being skeptical about finding a single quantity, whether conceived of as brightness or as an underlying signal that could exactly describe the results of all three methods. The maps of (brightness, L, L_b) space are attempts to see how far you can push that idea. They are enlightening for some purposes, but unlikely to be totally successful.

The claim I do want to make is rather for the importance of the more modest type of convergence. Such convergence is convincingly shown, and section 7 adds more evidence from other types of experiment, by other workers. Contrast brightness provides a coherent field, which we can explore by different methods. Both the agreements and the disagreements between the methods are informative, the former in showing robust common features, and the latter in warning us which features may be specific to particular methods or require further study.

6.2. Fechner's Integration

We have seen several correspondences between discrimination and brightness data. But are they sufficiently exact to support Fechner's hypothesis that brightness is the sum of JNDs? Yes and no.

The agreement with the scaling data is good, as shown in Fig. 2.17. The JNDs set in the scaling experiment were about 4 times larger than the discrimination thresholds, but the form of the relationship between $\Delta^2 L$ and L is the same. From one point of view this is not surprising: Both experiments can be seen as measuring JNDs. But the techniques differed in many ways (flashes versus steady lights, Maxwellian view versus CRT, psychophysical method, type of judgment, subjects, etc.), so the convergence against such odds considerably strengthens the common conclusions.

Some equivocal agreement with darkness matches is shown in Fig. 2.18. The points are the same matching data as in Fig. 2.8, but the curves are computed from the decrement discrimination JNDs. To do this I fitted curves to the discrimination data and derived a formula expressing the JND as a function of L and L_b, in much the same way as described earlier for isobrightness equations. Each curve in Fig. 2.18 is n JNDs above detection threshold, with n being chosen to give the best fit to each set of data points. Within the region where the discrimination threshold was measured, between the dashed lines, the agreement is quite good. In particular, the puzzling feature that the slopes are > 1 is reproduced in the JND

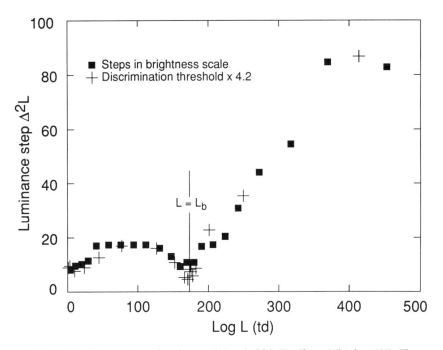

FIG. 2.17. Comparison of scaling and threshold JNDs (from Whittle, 1992). The squares show the luminance difference between adjacent circles in the brightness scale of Fig. 2.10. The crosses show discrimination thresholds like those in Fig. 2.14, except that $\text{Log} L_b = 2.35$ and these are the average data of the two subjects.

curves. This corresponds to the Weber fractions for discriminating decrements being slightly *greater* on brighter backgrounds, in both ΔL and L ranges (Fig. 2.14). This agreement is pleasing, but I doubt whether it is more than coincidence. Although the positive slope of the regression lines of the Weber fractions on $\text{Log} L_b$ gave this fit, the slope was not statistically different from zero. Nevertheless, Fig. 2.18 does demonstrate that a small increase in Weber fraction could, within a Fechnerian scheme, correspond to the fanning-out of the isodarkness contours.

For increments the relation between matching and discrimination is even less certain. The convergence of the isobrightness curves at high background luminances, so obvious in Fig. 2.2, which I described as higher exponential gain on bright backgrounds, is not reliably mirrored in the discrimination thresholds. Calculation shows that the gain difference in Fig. 2.2 between background luminances A and D should correspond, in a Fechnerian scheme, to discrimination threshold being 0.2 log units *less* on D than on A. A difference of about this magnitude was found by Whittle and Swanston (1974), and a hint of it can be seen in Thijssen and Vendrik (1971), who made similar measurements under rather different conditions. Such a difference is intuitively plausible, at least once one accepts that it is

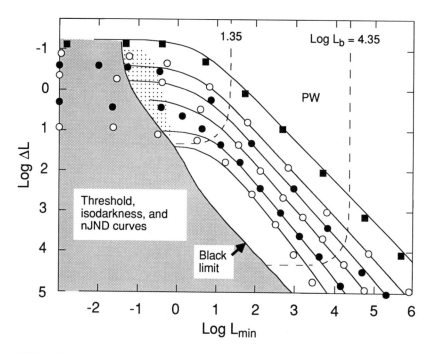

FIG. 2.18. An attempt at a Fechnerian integration. The points are the isodarkness data from Fig. 2.8. The curves are threshold and constant numbers of JNDs above it, computed from the discrimination thresholds of Whittle (1986), of which those in Fig. 2.14 are a sample. The lines showing $LogL_b$ values of 1.35 and 4.35 were the limits of the discrimination threshold measurements.

ΔL that is being discriminated rather than L, because the bright background makes the bright discriminanda less glaring. However, there was no sign of this difference in the more extensive measurements of Whittle (1986) made under conditions closely similar to those of the matching experiments.

This discrepancy remains as a puzzle for future work. It may simply be, for both increments and decrements, that the gain differences are more easily detected by matching than by measuring discrimination threshold. These thresholds are quite variable, cues other than brightness can sometimes be used, subjects can miss the most effective cue for long periods, the effect we are looking for is small, and the measurements of Whittle (1986) were not designed with the increment gain in mind. Also, the trends where they do exist go in the right direction. The question is given added interest by the suggestion raised in section 2.3.7 that a perceptual advantage of the increased exponential gain on bright backgrounds is to maintain the discriminability of bright flashes that would otherwise be masked by their halo of scattered light. The data show that discriminability is indeed maintained, but the link suggested would imply an actual improvement.

6.2.1. Brightness and Discriminability for Stimuli Seen Only with the S-Cones.

A more striking discrepancy between brightness matching and discrimination is provided by lights seen only with the S-cones. The S-cones contribute little or nothing to brightness (e.g., Eisner & MacLeod, 1980). One of the experiments in Whittle (1973) provided a nice example of this. Blue or violet flashes on a yellow background produced clear S-cone branches at threshold, where detection could be based on hue, but none at all in suprathreshold isobrightness curves formed by matching to a standard seen with the M- and L-cones. Now, the luminance discrimination of such lights, seen only by the S-cones, is poorer than for those seen by M- and L-cones (the Weber fraction is about 2.5 times greater; Whittle, 1974b), but if the discrimination were based on brightness, it ought to be impossible. Subjectively, it is based on something like salience.[30] This should warn us not to expect perfect accord between brightness matching and luminance discrimination.

6.3. Brightness Functions

The contour maps shown in section 4 represented only one out of three possible coordinate planes of the three-dimensional space whose axes are brightness, L and L_b. Figure 2.19 shows a different plane. These are *brightness functions*, showing how contrast brightness grows with luminance on different backgrounds. Brightness is again expressed as B/B_0, as in Fig. 2.11–2.13. The filled squares show the empirical equal-interval scale of Fig. 2.10, which was obtained at $\text{Log}L_b = 2.24$. The continuous curves were all derived from this one, by using Equation 4 to extrapolate it to a wider luminance range and the isobrightness equations transform it to other values of L_b.[31] This figure has two purposes. First, these brightness functions show some characteristics of contrast brightness particularly clearly (similar families of curves have been presented by Heinemann, 1961, 1972, and Heggelund, 1974):

1. Contrast brightness is all positive when $L_b = 0$ but can be negative

---

[30]Marks (1974) found that subjects could make magnitude estimates of the "brightness" of S-cone lights. Without the direct standard of comparison that a matching technique gives, it is possible that his subjects could have been judging something like salience, or brightness as in "bright colors." To investigate the apparent contradiction between our results, I repeated my heterochromatic brightness matching experiment on 10 subjects. Five of them found the task of matching a flash seen by the S-cones with one seen by M- and L-cones impossible. Four out of the remaining five gave results that showed wide fluctuations between different criteria. This and other results confirmed my original conclusion that S-cones make a negligible contribution to brightness as judged by heterochromatic brightness matching in the HSD.

[31]Equation 4 already expresses brightness as a function of ΔL and L, and hence L_b, but it is not suitable for directly generating a family of brightness functions because a and b also vary with L_b.

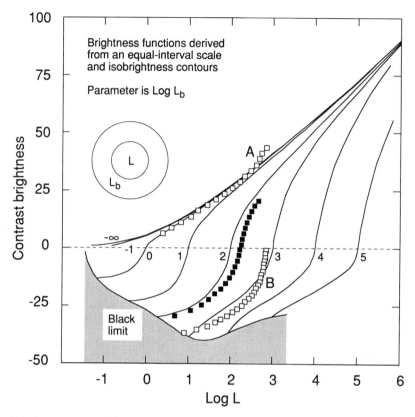

FIG. 2.19. Brightness functions showing how contrast brightness varies with luminance at different surround luminances. These curves are derived from the brightness scale of Fig. 2.10, transformed by the isobrightness equations that described the contours of Fig. 2.8.

when there is some surround light. This enlarges the range. (The range of negative values is largest when $\text{Log}L_b = 3$, which reflects the way the black limit varies with L_b but is probably misleadingly exaggerated in this figure because the effect of stray light in decrements has not been taken into account.)

2. Over substantial ranges it is approximately proportional to LogL.

3. For increments it grows more rapidly with LogL on bright backgrounds. This is the "higher exponential gain" discussed earlier.

4. The curves that cross the zero line are steepest at that point, where $L = L_b$. This is the Crispening Effect.

5. The curves converge at high values of Log L. L_b makes little difference there. This reflects the long plateaus in the higher isobrightness curves in Fig. 2.2 and 2.8.

6. Over a range of brightnesses that encompasses those we commonly see, say ±20, the curves differ mainly in their lateral position. This exemplifies the multiplicative gain model of adaptation (section 8).

The second purpose of this figure is to examine quantitatively the fit between the matching and scaling data. Equal-interval scales were obtained on backgrounds of different luminances (Whittle, 1992), though only the scale on a gray background has been used so far. Can we "predict" the scales on other backgrounds by transforming that scale with the help of the isobrightness functions? The two sets of hollow squares in Fig. 2.19 show attempts to do that. The upper scale, marked A, was for a series of increments on monitor black (0.04 log td), and the lower, B, a series of decrements on monitor white (2.86 log td). The procedure in plotting these points was first to calculate the brightnesses of the endpoints of the scale

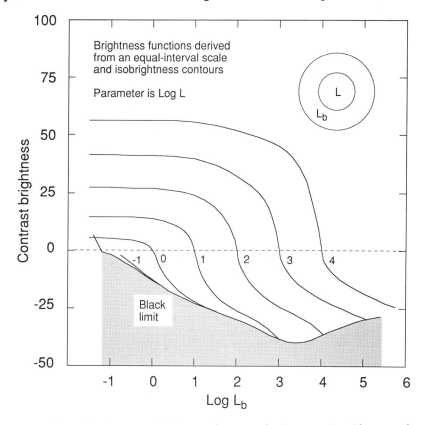

FIG. 2.20. How the contrast brightness of a constant luminance varies with surround luminance. These curves are derived from the brightness scale of Fig. 2.10, transformed by the isobrightness contours of Fig. 2.8.

by the same procedure used to generate the brightness functions (i.e., from Equation 4 plus the isobrightness functions). Then the brightness interval between the endpoints was simply divided into the appropriate number of equal intervals, because that was what the subjects were supposed to do, and the resulting brightness values plotted against the luminances the subjects actually set. This procedure amounts to placing the endpoints on the right brightness function, and then asking whether the intervening points also lie on the curve.

It works quite well for scale A, provided we omit the upper three points that are obviously divergent. But scale B is much more curved than the comparable brightness function (for $LogL_b$ = 3.0). I suspect that part of the problem is what also generates the implausibly deep dip in the black limit and the downward pointing tails on brightness functions 3, 4, and 5. This is that the extrapolation of the isodarkness functions into the region marked with a "?" in Fig. 2.8 may not be valid. This region represents almost-black stimuli. Satisfactory matches cannot be made there because you cannot bracket the desired value.

Figure 2.20 shows the third orthogonal plane: brightness versus Log L_b with LogL as parameter. The steep descent on the right-hand curves corresponds to the convergence of the isobrightness contours in Fig. 2.2 and 2.8. Note that for these curves more than half the total brightness loss as L_b is increased occurs while $L_b < L$. The generalization often made that simultaneous contrast is weak when $L_b < L$ finds no support in these data. Under other conditions it may reflect the qualitative change from increments to decrements.

The next two sections provide further arguments for favoring the weaker over the stronger form of convergence between the three experimental methods.

6.4. A Multiplicity of Contrast Signals?

The observation was briefly mentioned earlier (section 2.5) that isobrightness curves were not quite the same as iso-apparent-contrast ones. Shevell (1986) showed that they were also different from curves representing a constant contribution to a binocular brightness mixture (see section 9.3.2). These findings, which may be only two out of many possible similar ones, raise the suggestion that the eye transmits more than one contrast code to the brain. We should expect this from what we know of the diversity of cell types in the retina. Cells can be classified by several criteria: on versus off center, receptive field size, chromatic characteristics, temporal characteristics, and so on. For most of them response depends crucially on spatial contrast. But each may transmit different contrast expressions for the same stimulus. For instance, on and off center units near the edge of a test patch might do so, perhaps giving rise to the differences between increment and decrement isobrightness curves that we have seen.

The idea that a single edge may give rise to several different contrast expressions raises an interesting possibility. Although explicit representation of absolute luminance is lost in a contrast code, it may be recoverable at a higher level by comparing different codes. For example, suppose that all the expressions are of the form of Equation 3. Then, if the brain could somehow determine the value of the parameters n and L_0, it could in principle recover the absolute values of L and L_b from two local contrast signals. Such a scheme would predict absolute luminance perception in normal scenes but not in a stabilized image or a ganzfeld, which lack contrast signals. The arguments in the next chapter, that recovery of luminance information depends on more global factors, goes against this idea, but it remains a possibility that should be borne in mind.

6.5. Brightness as Just Another Contrast Dependent Response

Expressions of the form $\Delta L/(L_{dc} + L_0)^n$ are likely to describe most contrast dependent responses, because they capture two pervasive properties of sensory processes: sensitivity to small differences and normalization with respect to adaptation level. Therefore a pattern of isoresponse contours something like Fig. 2.8 should be obtainable from all such responses. We would then no longer see that graph as a representation specifically of brightness but rather as a particular standard kind of graph: a family of isoresponse contours. Like other standard graphs such as contrast- or spectral-sensitivity functions, it would prompt certain standard questions. Here are some examples. Some of these are questions that we are used to asking about single t.v.i. functions, and that context suggests possible answers also. For many of them we can look for answers at the levels both of retinal processes and of perceptual function (e.g., degree of constancy under illumination changes):

1. L_{dc} is whatever DC variable (L_b, L_{min}...) controls luminance gain. Are there responses for which L_b or the average luminance, L_{av}, are better than L_{min}? Note that if it were L_{av} as in Michelson contrast, there could be no plateaux, because as L_{av} tended to zero, so would ΔL.

2. How are the curves spaced out? The spacing must reflect the "response function" relating response to ΔL. For a response like frequency of seeing, which varies over only a small ΔL range, the response function saturates quickly and would produce two very narrow families of curves. Brightness or reaction time vary over the whole domain of visible lights

and so generate wide families like Fig. 2.8. Constant velocity or subjective contrast would be intermediate.

3. Are there responses that do not depend on a spatial difference ΔL, for which the numerator of the contrast expression would be L rather than ΔL? These would generate a family of curves in which L_0 decreased sharply above threshold.

4. How does the exponent n vary? Weber's Law is an achievement that the system does not always reach. For brightness, n is different for increments and decrements and varies in opposite directions as exposure duration is decreased. Why?

5. How does L_0 vary with response level? Most contrast expressions are going to have such a threshold or noise constant that has to be exceeded for L_{dc} to have any effect. Both L_0 and n can be seen as factors that make the response *depart from* exact dependence on contrast rather than merely as parameters in a contrast expression.

6. Extra branches may appear for additional receptor mechanisms, like the π' ones. In other heterochromatic or mesopic conditions, branches for other mechanisms would appear. If they do not, then why not?

I am trying to make two points by raising these rather obvious questions. First, I emphasize that brightness *can be seen* as just one of many contrast dependent visual responses. (If it is measured in certain ways: the qualification that lies behind the whole of this chapter.) I hope this too seems rather obvious by now, even though it is not a common point of view in the literature. Brightness has usually been treated as something special, perhaps because it has been seen as *the* primary subjective response to light intensity. Second, I point out that it would be interesting to have comparative measurements of different contrast dependent responses, and a general theory of both the mechanisms and advantages of contrast coding (e.g., Heeger, 1992; Koenderink, van de Grind, & Bouman, 1971; Ullman & Schechtman, 1982). That is the context needed for the preceding questions.

The next three sections deal with topics to which more than one of the experimental methods is relevant. They discuss substantive points that are of interest in various contexts and provide further illustrations of the way in which data from the different methods can complement one another.

6.6. What Produces Good Grays?

The decrement patches in the matching experiments looked like pieces of gray paper, except when their luminance was very low, when they looked black, with an overlying fog if L_b was also low (section 2.4.1). We are now in a better position to consider what stimulus conditions produce a good gray, and why. In Fig. 2.8, good grays are represented by all the decrement points outside the stippled regions; that is, by the descending parts of the isodarkness curves but not the initial plateaux. The transition comes not at the point where the light in the patch first becomes detectable (the "black limit"), but where it starts to have an adapting effect, which is between 0.5 and 1.0 log units above the black limit. This is shown by two indices. In Fig. 2.8 it is where the isodarkness curves start to descend. In the discrimination measurements of Fig. 2.14(c), the same value of L is where the value of $\Delta^2 L$ starts to increase. These facts are consistent with the idea that a good gray results from a decrement for which both luminances L and L_b are not just above detection threshold but are strong enough to have adapting effects. Because the gray seen depends on the luminance ratio, or other contrast expression, this accords with the idea (section 8) that the computation of ratios (contrast coding) is also the multiplicative component of adaptation.

Jung (1973) described a suggestive physiological correlate of this: "the essential condition for antagonistic field effects" (and so for contrast coding?) "is *not* general light adaptation but some *local illumination within the field centre*" (p. 34, emphasis in original); that is, the receptive field center has to be sufficiently illuminated to have an adapting effect.

6.7. The Usefulness of $W = \Delta L / L_{min}$

This expression, or a close relation of it, has appeared several times. What makes W an unusual contrast expression is that it uses L_{min} in the denominator rather than the more usual L_b. This makes a difference only for the higher contrast decrements, approximately those for which $\Delta L > L_b/2$ (Whittle, 1986). For increments $L_{min} = L_b$ and for low-contrast decrements, the difference is insignificant. As was pointed out earlier, the advantage of using L_{min} is that it captures the sensitive discrimination of the amount of light in higher contrast decrements. This sensitivity shows up in all three types of data.[32]

It is showing up in other situations too. In preliminary experiments, Kingdom and I recently found that contrast discrimination of 0.5 cycle/degree sine- or square-wave gratings is made on the basis of L_{min} at high contrasts. This agrees with the common report that subjectively such discriminations are based on the darkness of the dark bars. Recent experiments on the matching of contrast colors (see section 7.2.3) by Shepherd, in

---

[32]The following note from Whittle (1986) may be helpful here. "Visual science requires convenient expressions for relative luminance, that is, contrast. ΔL and R [$= L_{max}/L_{min}$] are the simplest. However, ΔL makes no reference to Weber's law, and R makes none to our sensitivity to small differences. $\Delta L/L_b$ masks the full extent of increment–decrement symmetry. Any function of R expresses that symmetry, and W and M combine all three merits. M is appropriate for linear analysis but fails to represent our good discrimination of high contrasts. W does the reverse" (p. 1688).

Cambridge, suggest that the equivalent of L_{min} for each cone fundamental has the same advantage that L_{min} had for describing darkness matches. Finally, Carter (1993) found that W gives a good account of a performance measure of visual search for a trapezium of a specified gray among a large array of distractor rectangles of various positive and negative contrasts. This experiment was done by Williams (1966), who recorded the frequency with which subjects fixated on the distractor rectangles and so derived a measure of contrast confusion, and therefore of discrimination. The sensitivity to L in higher contrast decrements thus seems to be a robust result.

W is mathematically convenient for summarizing some aspects of brightness and contrast perception. It must therefore at some level summarize the underlying physiological processes, but the question is, at what level? Is it possible, for instance, that in the retina multiplicative gain for increments is set by L_b and that for decrements by L? I have not found relevant physiological data. Kingdom and Moulden (1991) argued that such asymmetry in gain-setting was implausible and proposed instead an account of the luminance discrimination data of Whittle (1986) based on the quantity $G = \ln(L/L_b)$. Functional interpretations of W are a large topic that cannot be dealt with here, but it is relevant to the theme of this chapter to point out one advantage of W as a possible descriptor of retinal processes. This is that, like Michelson contrast and unlike G or the usual Weber fraction $\Delta L/L_b$, it is defined in terms of L_{min} and L_{max} rather than L and L_b. Now the distinction between L_{min} and L_{max} is available to any local receptive field or edge detector, whereas that between L and L_b is not. (In the insets to Fig. 2.2 the homologous roles of ΔL and L_{min} in increments and decrements are indicated by the heavy outline.) The latter distinction, for steady lights, expresses which is figure and which is ground, and these are defined only globally. Therefore no function that depends on that distinction can, for purely logical reasons, describe the behavior of local detectors.

6.8. Summary of Increment–Decrement Differences

It may be useful to summarize these in one place. The similarities are shown in the symmetries, striking even though partial, that can be seen in different ways for all three types of measurement, matching, scaling, and discrimination, in Fig. 2.2, 2.7, 2.8, 2.10, and 2.14.

Decrements show the following differences from increments. Items (1) to (6) are true in the HSD. They probably also hold in several related conditions (see next chapter), but we do not know to what extent:

1. They always look darker.

2. They appear as surface colors rather than self-luminous, under the present conditions. With less sharp edges decrements look like shadows.

3. It is not always possible to make satisfactory matches (the stippled regions in Fig. 2.8). Specifically, matching is difficult when the patches are seen as black.

4. The slope of suprathreshold isodarkness curves (n in Equation 3) is >1 and increases with darkness level. This implies that for a given decrement contrast there is an optimum absolute luminance level at which it looks darkest, whereas increments of fixed contrast reach their maximum brightness at a certain absolute level and then maintain it.

5. Suprathreshold isodarkness curves do not shift to the right relative to threshold nor show an upper branch (Fig. 2.2 and 2.8). Appearance is a function only of contrast over a longer range than for increments, except for the qualification in (4).

6. Lengthening flash duration has opposite effects: It increases the exponential gain of increment brightness with respect to ΔL (section 2.3.1) on bright backgrounds but decreases that of decrement darkness.

7. Above detection threshold, discrimination threshold is an inverted-U-shaped function of ΔL for decrements, whereas for increments it increases monotonically (Fig. 2.14).

8. Correspondingly, an equal-interval lightness scale for decrements is inflected, with a shallower slope in the center than at either end (Fig. 2.10).

Similarities and differences are always relative to a particular representation. Quantitative differences, like (7) and (8), can be made to appear formally as similarities by appropriate mathematical transformations. I have done this in several cases, in both graphs and equations, by substituting L_{min} for L_b. Another interesting case is the slope asymmetry in (4). It appears as a difference when the ordinate is LogΔL, but a similarity when the ordinate is Log$L$, as in Fig. 2.11. What is the right level of explanation for it?

The perception of increments and decrements, or brightness and darkness, has commonly been attributed to on and off center neurons (e.g., Jung, 1973). This attribution does not usually rest on detailed correspondence between psychophysics and physiology. A list of contrasting characteristics like this one could be of use in strengthening the argument. Specifically, do items (4) to (7) also differentiate between on and off center units?

6.9. Sources of Misunderstanding

Before leaving my own experiments, I want to comment briefly on some features of them that I have found to cause misunderstanding. I am looking over my shoulder at a number of different audiences. The conceptual frameworks in which such an apparently straightforward topic as brightness has been discussed over the last hundred years have been, and still are, amazingly diverse. Compare, for example, the systems analysis approach of Du Buf (1987) with the phenomenological one of Heggelund (1991) or more traditional treatments to be found in MacLeod and Pick (1974). Many workers in this field must see my approach as wrong-headed in various ways, disregarding facts and distinctions that seem to them obvious. The most important reason for this is my presumption that vision is based on seeing spatiotemporal contrast. Here are some points of potential misunderstanding. Items (2), (4), and (6) are discussed at more length by Whittle and Challands (1969), and (2), (3), and (5) in the next chapter:

1. Uniform sharp-edged objects are used (step or bar profiles as opposed to sinusoidal or Gaussian ones) because they are common in natural scenes, and to link up with previous work that has used them. I hope that the present results on contrast brightness may help to bridge the gap between old and new psychophysics, but this is a large task that has scarcely been started.

2. The choice of the HSD is defended in section 2.1, in Whittle and Challands (1969), and implicitly in section 5 of the next chapter. Its main disadvantage is the possibility of binocular interaction because each patch is seen through the contralateral surround. I believe that such interactions are minimal (Whittle & Challands, 1969), and that they are a price worth paying for the ease of matching and reliability of data that the HSD allows. It is not sufficiently appreciated that they can be avoided only at the cost of opening the experiment to other high-level influences that are in practice more troublesome (see next chapter). There is no perfect solution and the truth will emerge only from converging operations. Finally, it is a recurrent argument in these chapters that conclusions drawn from the HSD are confirmed by measurements in situations in which such binocular interactions are absent.

3. Reduced and artificial conditions are used for the usual reasons in psychophysics: to produce mathematically simple results, and because of an interest in mechanism. The disadvantage is that the relevance to normal vision is uncertain.

4. Confounding simultaneous and successive contrast. A flash on a uniform background is one spatiotemporal transient. To isolate "brightness contrast" and "adaptation" requires introducing additional transients that complicate the situation and make the results harder to analyze.

5. Lumping together lightness and brightness. They are not confused but compared as decrements and increments.

6. There is a recurrent ambiguity as to whether the perception of contrast or luminance is in question. This is discussed later in section 10.1. The point is that in these experiments brightness is *determined by* contrast.

7. TWO MORE TYPES OF EVIDENCE

7.1. Dynamic Contrast Brightness

Contrast brightness is dramatically shown in a demonstration like the one described in section 1.1 in which the luminance of either test patch or surround can be modulated in time. The light in the patch appears to be waxing and waning regardless of whether the physical modulation is in patch or surround. Both kinds of modulation have been studied in a series of papers by Glad and Magnussen and colleagues. They used a haploscopic display like Fig. 2.1 but with the surrounds seen one above the other in the binocular field. Subjects set a flash seen by one eye to match the maximum or minimum brightness of the modulated patch seen by the other. Their results seem to show the same separation of increments and decrements as in Fig. 2.2, even when the surrounds were of very different luminances in the two eyes (Magnussen & Glad, 1975a). There was also a strong dynamic Crispening Effect, in that the difference between maximum and minimum brightness for a given luminance modulation was greatest when the surround luminance was within the modulation range and fell off steeply outside it.

The separation of increments and decrements is strikingly confirmed by a reanalysis of one of their experiments. The matches to maximum brightness and darkness were markedly asymmetrical about the mean brightness. This was so even when brightness was plotted on a subjective scale, of the same form as that in Fig. 2.10, obtained by summing JNDs (Glad, Magnussen , & Engvik, 1976). The asymmetry was such that the light phase of 85% modulation produced about the same departure from average brightness as the dark phase of 45% modulation. However, precisely this degree of asymmetry is to be expected if increments and decrements are processed separately, a principle that was clearly stated by Björklund and Magnussen (1979): "There is increasing evidence that incremental and decremental aspects of spatio-temporal luminance distributions are processed by separate mechanisms" (p. 157). This is because a symmetrical square-wave modulation can be regarded as a train of increment pulses plus a train of decrement pulses 180° out of (temporal) phase. The increment half of 85% flicker has a spatial Michelson contrast of 30% with respect to the steady surround. The decrement half of 45% flicker has a contrast of 29%. Therefore the results of Glad et al. showed that the fluctuations in brightness and darkness around the mean were equal when the spatial Michelson contrasts of the increment and decrement trains *taken separately* were equal. This result is in approximate agreement with the contour map of Fig. 2.13 at the background luminance used by Glad et al. Note that the decomposition into increment and decrement trains makes little sense in purely temporal terms, reinforcing the arguments that follow (section 9.1) that it is primarily the *spatial* contrast that determines contrast brightness.

The separation of increments and decrements also accords with another aspect of the appearance of a patch flickering between positive and negative contrast in a steady surround. At around 8 to 10 Hz, where Magnussen and Glad found the greatest brightness or darkness enhancement, either the bright or the dark phase often almost disappears for 2 or 3 seconds, leaving a black square with only slight flicks toward light, or else a light one with only slight dark flicks (Whittle, 1963). This "monocular rivalry" looks quite like the binocular rivalry that occurs when increments and decrements are presented to opposite eyes. It becomes more comprehensible if the increments and decrements are processed by separate channels, as in other cases of monocular rivalry (e.g., Campbell & Howell, 1972).

The experiments of Glad, Magnussen, and colleagues thus confirm by their different methods not only the basic contrast dependence of brightness but three of the specific conclusions concerning contrast brightness: the separation of increments and decrements (confirmed in three different ways), the existence of the Crispening Effect, and the predominant importance of spatial contrast as opposed to temporal.

De Valois, Webster, De Valois, and Lingelbach (1986) also compared the effects of modulating test patch and surround. They found that induced brightness would follow surround modulation only up to a frequency of 2.5 Hz and argued that this sluggishness of induction suggests a cortical locus for it. Magnussen and Glad (1975b) had done the same experiment at much higher luminances and found following at somewhat higher frequencies (6 to 10 Hz), but they too found that induced brightness fluctuations could not be driven as fast as those produced by directly modulating the test patch. This is to be expected from the two-stage model of simultaneous brightness contrast suggested later, in which the second stage is a relatively slow cortical process of filling-in that could not follow the higher flicker frequencies. This suggestion would also be consistent with the observation of De Valois et al. (1986) that at high frequencies "the induced pattern stays a constant gray but nonetheless continues to flicker very strongly, particularly around the edges" (p. 891).

7.2. Contrast Color

7.2.1. Existing Evidence. Much of what has been said about contrast brightness can be generalized to color. Under conditions where brightness is a function primarily of local luminance contrast, color is a function primarily of local color contrast. I use the term *contrast color* as a generalization of contrast brightness, and for the same reason to remind us that we are talking about something slightly special. It refers to situations where local contrast is a dominant variable, and what is true of such situations should not be assumed to be true generally. Hering introduced the HSD in 1890 because it demonstrated particularly strong contrast colors. Evans (1974) discussed at length the striking contrast colors produced in center surround displays. Mausfeld and Niederée (1993) offered an ambitious theoretical treatment.

Land's two-color projection demonstrations, and the responses to them by Judd (1960) and Walls (1960) and Wilson and Brocklebank (1960) in terms of the known facts about chromatic adaptation and simultaneous contrast, suggested how many aspects of color appearance may be understood in terms of contrast color. There was also a remarkable and neglected paper by Spencer (1943) proposing a very simple model of contrast color in terms of the vector between the points representing patch and surround colors in RGB space, which accounted with elegance and quantitative precision for Helson's (1938) classic results on the effects of illumination on perceived color. Bergström's rather different vector model (chap. 6) has related implications.

Some published work on color is closely related to the experiments I have described on brightness. Whittle (1973) reported hue and saturation judgments in the HSD for red and green increments superimposed on red and green backgrounds, in all four combinations. Physical color mixture or any psychological additive effect of the background color on that of the patch would make a red patch, for example, look more red on a red background than on a green one. What happened was the opposite: A red patch looked more red on a green background than on a red one. For color, just as for brightness, the dominant effect of the background was its adaptive attenuation of superimposed increments.

Walraven (1976) showed this more precisely by a hue cancellation technique. An increment composed of a mixture of red and green was presented on a red background, and the subject adjusted the red–green ratio to make the mixture look yellow: neither red nor green. The more intense the red background, the greater the proportion of red needed. Walraven coined the phrase "discounting the background" to describe the fact that, although the red light in the background added physically to the mixture, this addition was somehow discounted, leaving only its adapting effect.[33] Isoresponse contours for the red component, where the constant response was to cancel the hue of a fixed green increment, had a similar

---

[33]"Discounting the background" is to be contrasted with "discounting the illuminant," which is a traditional description of color constancy. The former is subtractive, and the latter multiplicative discounting. It is noteworthy that in his first experiments Walraven presented the test lights in a hole in a red surround, rather than superimposed on a uniform background. It was the form of the data that made him realize that the visual system was treating the test lights as increments on a background.

form to the increment isobrightness contours in Fig. 2.2, showing a similar rightward shift (though their shape was invariant). There has been controversy as to how complete the discounting is in the chromatic case (Shevell, 1980; Walraven, 1981). But although that question remains somewhat open, Walraven's work shows that contrast color behaves very like contrast brightness. The correspondence was shown in more depth in later work using an analysis in terms of Stiles' π mechanisms, like that used by Whittle (1973) for brightness. Walraven (1981) concluded that: "The results can be described by assuming non-additivity of test and adapting light, cone spectral sensitivities as estimated by Vos and Walraven (1970), and receptor specific gain controls that have the same action spectra as Stiles π mechanisms" (p. 611).

7.2.2. Cone Contrasts. Walraven's conclusion outlines a specific model underlying the generalization of contrast brightness to color. The fundamental idea is that perceived color depends on the relative physical contrasts stimulating each of the three cone mechanisms separately. These are the *cone contrasts* that have been used to good effect by, for instance, Stromeyer, Cole, and Kronauer (1985), to distinguish between adaptational processes in the trichromatic and opponent stages of the visual pathway, and by Walraven, Benzschawel, Rogowitz, and Lucassen (1991) to describe color constancy. It has been realized for some time (e.g., Alpern, 1964, and one could trace the idea back to Young's sympathetic theory of color contrast in 1807) that such a scheme qualitatively explains the familiar properties of color contrast. Figure 2.21 shows an example of how hue could depend on the relative cone contrasts, with sign being taken into account. Thus a positive L-cone contrast would contribute redness and a

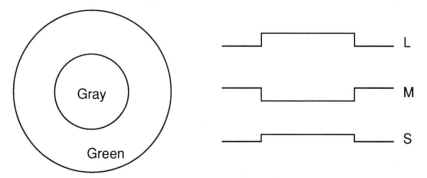

FIG. 2.21. Explaining color contrast in terms of the separate cone contrasts. L, M, and S show hypothetical response profiles in Long, Medium, and Shortwave cones to a gray patch in an equiluminous green surround. The positive L and S contrasts and the negative M would be expected to combine to make the gray look purplish, as it in fact does.

negative one would contribute green–blueness. A similar scheme of contrast coding in three chromatic channels underlies Land's retinex theory (Land & McCann, 1971).

7.2.3. Some New Measurements. All the methods described in this chapter for studying contrast brightness can be applied to color. Arend and I recently did some color matching in the HSD (Whittle & Arend, 1991). With control over the color of the test patch as well as its luminance, it was possible to make very satisfactory matches, and the two subjects in our experiment agreed extremely closely.

We used standard stimuli that had only a luminance difference between patch and surround (homochromatic), because for such stimuli the idea that hue depends on the relative cone contrasts makes the unlikely predic-

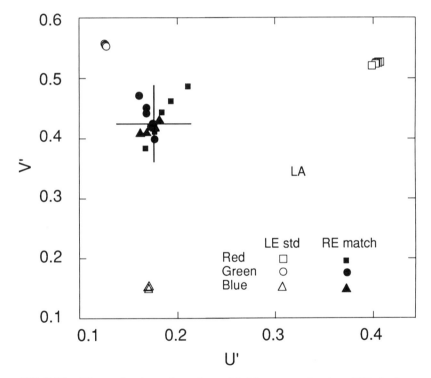

FIG. 2.22. The results of complete color-and-brightness matches in an HSD display like Fig. 2.1. The standard stimuli (hollow symbols) were homochromatic red, green, or blue displays with luminance contrasts varying between +30% and –30%. The filled symbols show the matches made by one subject for a variable patch in a gray surround of the chromaticity shown by the large cross. A second subject agreed closely. This is a CIE u'v' chromaticity diagram, so brightness is not represented. Stimuli were steady lights, and fixation was not constrained.

tion that they should all look the same color, presumably achromatic, whatever their physical color. This is because the ratios of the cone contrasts are always 1:1:1, because all three are equal to the luminance contrast. We knew that such stimuli did not usually look achromatic, but we thought that some might do so under contrast color conditions. This turned out to be true in the HSD. Figure 2.22 shows the matches set in a gray surround to saturated red, green, and blue homochromatic stimuli with Michelson contrasts from +30% to –30%. The center of the large cross marks the chromaticity of the gray surround, and the matches cluster around it. Matches to the higher contrast increments were desaturated versions of the standard color, and to the higher contrast decrements were approximately complementary to it. This is an instance of the Helson–Judd effect for surface colors in chromatic illumination (e.g., Evans, 1974). This kind of experiment may throw some light on the mechanism of the effect. It was the low-contrast (7%) decrements in each case that were truly achromatic. When the data were plotted in terms of cone contrasts, calculated from the fundamentals of Smith and Pokorny (Wyszecki & Stiles, 1982), it could be seen that many matches followed the rule that the left- and right-eye contrasts were equated for each of the three fundamentals. Some of the deviations could be accommodated by the use of a ratio like that in Equation 3 rather than simple contrast.

7.2.4. Conclusions on Contrast Color.

There is therefore already much evidence for the old hypothesis that hue depends on physical color contrast. There are good reasons to believe that this idea has the power to radically change our understanding of color appearance, and to organize much of the huge literature on chromatic adaptation, contrast, and constancy (see the references given earlier in section 7.2.1). But if it has such potential, why has it not found more general acceptance? There are three obvious reasons. First, much color science has concentrated on problems of metamerism that concern the very first stage of the visual pathway—the photopigments—where contrast colors are irrelevant. Second, just as one of the main arguments of this chapter and the next is that the relation of brightness to luminance contrast is very clear in certain situations but totally obscured in others, so the relation of hue to physical color contrast is clear only in particular situations. Other situations provide such striking counterexamples that the fundamental nature of contrast colors can be consistently maintained only as part of a discouragingly complex theory. Nevertheless, that may be what is required to understand color appearance (as many color scientists have always feared). The third reason is that there are some thought-provoking differences between brightness and hue. If contrast colors depend on the relative cone contrasts, then they are severely constrained by the overlap in spectral sensitivities of the L and M cones. Another difference, or at least a puzzle, is the lack of low-frequency cutoff in the chromatic contrast sensitivity function (Mullen, 1985). This suggests that edge coding should be less marked for color than for brightness.

In spite of such problems, it seems possible that contrast colors will provide not so much an extension of the brightness work as a strengthened foundation for it, for two reasons, one empirical and one theoretical. The empirical reason is that more satisfactory matches can be made when one can vary color as well as brightness. Even in the supposedly "achromatic" conditions of Fig. 2.2, there was often a hue difference between the patches, particularly for decrements, and sometimes it was subjectively troublesome. In color matching the hue difference is removed as part of the task, so the data should be even more reliable than they are for brightness, and our results so far confirm this. The theoretical reason is that, because brightness must depend on some combination of separate cone channels, it should also be explicable in terms of them.

8. SENSORY PROCESSES?

8.1. Contrast Brightness and Retinal Processes

There are several reasons for believing that brightness matching in the HSD reflects retinal processes: principally, that curves of the form of Equation 1 can be produced by physiological measurements of retinal ganglion cells, and that the effect of the background in the HSD depended on its monocular luminance rather than its brightness in the binocular field.[34] This argument is strengthened by the comparison with an effect of background brightness (assimilation), that does indeed depend on the brightness of the background in the binocular field (Shevell, Holliday, & Whittle, 1992; see next chapter), and not its monocular luminance. The question of a possible role for brain processes in determining contrast brightness is taken up in section 9.3.

The matching data are described by Equations 1 and 3, or variants of them, and the scaling and discrimination data by functions of W. All these expressions involve a ΔL term divided by a function of luminance. This suggests that all the measurements reflect two basic sensory processes: differencing and multiplicative attenuation. The retina computes ΔL, probably at edges, and attenuates the signal by a function of L_b or L_{min}.

---

[34]An important incidental observation is that even when in the HSD the appearance of the binocular background fluctuates in brightness or color or both, because of binocular rivalry, the appearance of the patches is not affected at all. This is what would be expected if it depended on the *physical* backgrounds, not their appearance.

The idea of a multiplicative gain mechanism of adaptation is well established. The subtractive component is slightly less familiar, but a model of adaptation involving both components has been evolved over the last 15 years by Geisler, Williams, Adelson, Hood, Hayhoe, and others. A useful recent treatment was that by Hayhoe, Benimoff, and Hood (1987). Walraven, Enroth-Cugell, Hood, MacLeod, and Schnapf (1990) also provided a convenient summary. The same two components are involved in the coding of physical contrast. "Rapid adaptation" and "contrast coding" probably refer to closely related retinal processes, which are reflected also in the behavior of contrast brightness.[35] The next sections discuss attenuation and differencing separately. There is a great deal of evidence that both occur, though the main psychophysical justification for reifying them as distinct retinal processes comes from the work just referred to.

8.2. Attenuation

This is most familiar as Weber's Law, or in current jargon a multiplicative gain control with attenuation factor proportional to background luminance. My data accord with this where isobrightness curves have slope 1.0. The range can be extended by adding the "dark light" constant L_0 to L_b, as in Equations 1 and 3.

I say no more about this here. The interested reader should consult Walraven et al. (1990) and Shapley and Enroth-Cugell's (1984) review on gain-controls. There is of course a huge weight of evidence for Weber's Law as an approximate description of light adaptation under many, though not all, conditions. Its mechanism remains obscure, so that it is still possible for radically new theories to be proposed, such as that of Laming (1986). It is fundamental to the theme of this book: the dependence of brightness and lightness on relative luminance. Its relation to lightness constancy is discussed in the next chapter and elsewhere in the book.

8.3. Differencing

All sensory systems have evolved to be exquisitely sensitive to small deviations around their momentary adaptation level. One mechanism underlying this is the computation of intensity differences around that level. In vision this means that ΔL is the effective stimulus in that zone. We

---

[35]The term *adaptation* has been used with various meanings. I mean here adjustments of sensitivity, latency, and so on, produced by changing retinal illumination, whether slow, fast, or virtually instantaneous. Functionally, this means those retinal processes that allow the eye to function over a wide range of light intensities using neural units that have only a restricted dynamic range.

infer this from many different types of evidence. For example, physiological responses such as evoked potentials or changes in ganglion cell firing rate are proportional to ΔL for small values (typically up to a Michelson contrast between 10% and 20%). Most work with spatial or temporal sinusoids depends on, and takes for granted, the functional significance of ΔL. Many results in the previous sections, such as the analysis in terms of Stiles' π mechanisms, or the demonstration of Bloch's Law for ΔL on a bright background, depended on the stimuli being physically composed of increment and background (ΔL and L_b). But however the stimuli are physically produced, the analysis into increment and background is often needed to explain the results.

The significance of ΔL in these experiments is most concisely shown by the fact that a ΔL term is needed in the equations and the graphs that best describe all three types of data. It is at low contrasts that it makes the difference. At high contrasts ΔL is effectively equal to L for increments and L_b for decrements. For the brightness matching experiments the choice of ΔL as a variable was important, and unusual in 1969 in studies of brightness. It was only the use of LogΔL that showed the similarity of isobrightness and threshold curves. In the contour map in which ΔL is not represented, Fig. 2.11, all one can see along the diagonal is the clustering of contours that represents the Crispening Effect. This effect appeared in both the scaling and discrimination data, perhaps most dramatically in the deep pointed notch at $L = L_b$ in Fig. 2.14(c). The mathematical form taken by this enhanced discriminability around adaptation level shows that it is produced by the eye's sensitivity to ΔL. In this range discriminability follows Weber's Law with respect to ΔL, not L. Similarly, Fig. 2.5 showed that Bloch's law of temporal integration applied to ΔL, not L. In these processes the background appears to be completely discounted.

Increments were always brighter than decrements. This is how it should be if the effective stimulus is ΔL, and the visual system preserves its sign, perhaps by coding increments and decrements separately. The increment–decrement separation does not by itself imply the coding of luminance differences. It could equally well be the consequence of separate coding of luminance ratios greater or less than unity, though that seems a less likely boundary. But in fact it seems to coexist with the metric properties just discussed. I suggest that there may be a very general principle involved, which can be formulated as follows (Whittle, 1992). It is a generalization of the Crispening Effect.

Let I be a stimulus dimension, with a reference level I_b, such that an object or event is *defined* by a differential magnitude $\Delta I = I - I_b$, in the sense that:

1. The object disappears when $\Delta I = 0$, and

2. Some subjective dimension, S, of the object is a monotonic function of the magnitude of ΔI. (This implies "increments brighter than decrements." It may be an invariable accompaniment of (1).) Then for values of I around I_b three things occur:
3. Enhanced discrimination between two such objects, perhaps following the Weber Law $\Delta(\Delta I)/\Delta I$ = constant.
4. Increased rate of change of appearance, $\Delta S/\Delta I$.
5. Simultaneous or successive contrast effects.

This applies to contrast brightness, and probably to contrast color. Other promising candidates are perceived depth produced by either stereoscopic or velocity disparities, and apparent contrast. The existence of contrast effects in these dimensions (Chubb, Sperling, & Solomon, 1989; Graham & Rogers, 1982) encourages this line of thought. As Chubb et al. (1989) suggested, the general theory of contrast coding, or normalization, which was called for in section 6.5, may be relevant to many stimulus dimensions.

Although the separation of increments and decrements now seems natural and obvious to most psychophysicists and physiologists, in the context of brightness it does have the still surprising implication of complete discounting. Not only is the background *luminance* completely discounted in the discrimination of superimposed increments but also its *brightness* contributes nothing to their brightness. The luminance of the background is important for its adapting effect, but, however great a pedestal of brightness it would seem to provide for a superimposed patch, the brightness of that patch is not increased in the slightest. If it were, a light at threshold on an intense background should look brighter than one in the dark. The threshold curves could not be isobrightness contours. But they are. This is a striking instance of the removal of the common component that Bergström discusses in chapter 6.[36] The intransitivity of brightness relations that it implies is discussed in the next chapter.

The functional significance of ΔL is therefore a very firm conclusion, because so many different types of evidence converge on it. However, those studying lightness and brightness, including the other authors in this book, continue to describe and explain their results in terms of luminance, L, albeit relative to L_b, rather than differential luminance, ΔL. Although I think this sometimes leads them to miss important points, there are never-

---

[36]However, it does not depend on the common component being perceived as illumination (Bergström's "ilco"). Indeed the background does not have to be perceived at all. This is retinal preprocessing, as it were, appropriate to Bergström's hypothesis under some conditions but not under others. In the latter, it may have to be undone at higher levels in the visual system.

theless good reasons for their recalcitrance. First, ΔL is not forced on your attention if you avoid working near threshold. Studies of lightness have often used stimuli such as gray papers, or discs in separately projected rings, that prevent work near threshold. To do that, zero-contrast stimuli must be available that are really invisible, such as are provided by CRT displays or by superimposed lights. These are special cases. Second, even though ΔL is encoded in the retina, the brain can decode, or integrate, in such a way that our perception is sometimes more simply related to L than to ΔL. This too is discussed in chapter 3.

9. EXPLAINING CONTRAST BRIGHTNESS

9.1. Contrast Brightness Depends on Spatial, not Temporal, Differences

Differences can be taken across space or time, and we know that the visual system handles them in some ways very similarly (Robson, 1966). Nevertheless, it is spatial differences that usually define the objects of visual perception, and this was the case for all the measurements of contrast brightness described in this chapter. The objects being judged were *defined* by edges, and this was essential both to the appearance and to the quantitative results.

The flashing stimuli in the matching experiments vary in the same way over space and time: a bar profile as in Fig. 2.1(b) in both dimensions. There has to be some temporal variation or else you see nothing. But the asymmetry is that the objects to be matched, the test patches, must be differentiated from the background by spatial, not temporal, steps. Some evidence for this has already been given in section 2.6 (flashing surrounds instead of patches) and section 7.1 (dynamic contrast brightness). We can also infer the importance of spatial differences by imagining various experiments in which the spatial or temporal steps *that differentiate patch from background* are removed.[37] To remove the temporal steps but not the spatial ones requires only a minor change in procedure. Rather than using flashes on steady backgrounds, switch patches and backgrounds on and off together. The same judgments can still be made and results of the general form of Fig. 2.2 are still produced. Now do the opposite: Keep the temporal steps but remove the spatial by making patches and backgrounds the same size.

---

[37]I have done some of these experiments, more or less formally, but do not take up more space here by citing detailed but untidy evidence. It is clear enough approximately how they would come out, just from thinking about them, informed by data elsewhere in the chapter.

This radically changes either the appearance or the results. If the patches are expanded to the size of the backgrounds, you cannot do the experiment at all because the patches are indistinguishable in the HSD. If the backgrounds are shrunk to the size of the patches, then the task becomes the quite different one of comparing brief blips in the brightnesses of two steadily visible patches. The blips are events rather than objects and do not "have brightness" unless they are very intense. One could match their salience and might produce results quite like Fig. 2.2, but the results would not describe brightness.[38] If the flashes lasted a second or more, you could match the brightnesses of the combined flash plus background, but the results would be very different: As duration increased the matches would tend toward equating total luminance, not contrast. Background luminance could no longer be discounted because there would be no basis for distinguishing it from patch luminance. It would simply add in.

Similar arguments can be made for the discrimination and scaling experiments. It is significant that they are most convincing for low contrasts, which is also where ΔL is necessary in describing the data from the main experiments.

I conclude from this that if I am right in distinguishing a syndrome, as it were, of contrast brightness that is a mode of appearance that occurs in certain situations and that obeys certain quantitative rules as key variables are changed, then it depends primarily on spatial differences. Because, if the spatial difference is removed, the appearance or the quantitative results or both are radically changed, whereas this is not so if the temporal difference is removed. You can find interesting similarities in the former case, but you have broken up the syndrome of contrast brightness and are talking about something else.

9.2. Is Discounting Explained by a Subtractive Component of Light Adaptation?

Hayhoe et al. (1987) suggested that discounting the background brightness could be explained by a subtractive component of light adaptation. However, within their analysis there are two ways to do this, and the difference between them raises fundamental questions in the explanation of contrast brightness. To make this clear, it is necessary first to give a brief exposition of their model.

They assumed, in agreement with much physiological evidence, that

---

[38]This would become an interesting experiment if the backgrounds were stabilized. The general question it would attack is how the contrast response to an edge is affected by asymmetrical adaptation strictly restricted to one side of the edge.

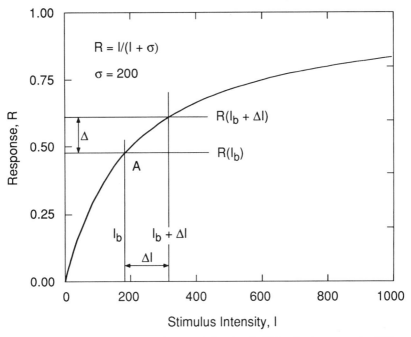

FIG. 2.23. Response-intensity function following the Naka–Rushton equation $R(I) = I^n/(I^n + s^n)$ with $n = 1$ and $s = 200$.

the response, $R(I)$, of retinal cells to stimulus intensity I follows the Naka–Rushton equation (Fig. 2.23):

$$R(I) = \frac{I^n}{I^n + s^n} \quad (5)$$

They further assume that an increment ΔI will be seen if it increases R by more than some criterion amount, Δ. Therefore an increment ΔI added to a background I_b will reach threshold when

$$R(I_t + \Delta I) - R(I_t) = \Delta \quad (6)$$

Because (5) is a negatively accelerated function, Equation 6 implies that the ΔI required to produce a constant response increment Δ will become infinite when I_b is large. For cones, such "saturation" occurs for flashed backgrounds but not for steady ones. This difference is explained in the model by supposing that light adaptation to a steady background modifies the inputs to $R(I)$ in such a way that (6) becomes

$$R(mI_t - x + m\Delta I) - R(mI_t - x) = \Delta \quad (7)$$

where m and x represent multiplicative and subtractive components of adaptation. After a change in background luminance, m and x take up values such that the response to the background, $R(mI_b - x)$, is reduced to zero, so that in the figure the point A effectively returns to the origin and saturation is prevented. The experiments of Hayhoe et al., and the earlier work they built on, provided strong empirical support for this model. They showed that both multiplicative and subtractive terms are required and measured the time-courses over which m and x change at both onset and offset of the background I_b. m acts faster, and probably more distally. The model gives a good account of much data on luminance increment threshold and on color appearance (Hayhoe & Wenderoth, 1991).

Hayhoe et al. (1987) suggested that the subtractive component represented by the $-x$ in Equation 7 also explains discounting. This is at first sight an attractive idea. The subtractive term reduces the background response to zero, and discounting seems to imply the same. However, on closer inspection it turns out that the term *discounting the background* is being used in two different senses, corresponding to discounting on the one hand the brightness of the background and on the other hand its adaptive effect. These have opposite consequences.

In my use of the term, if the *brightness* of the background were not discounted, then it would add to the brightness of an increment, so that *less* incremental light would be required to achieve a given brightness. Whereas without the subtractive component of *adaptation* (the $-x$ term), *more* incremental light is required for the response difference to reach Δ. Empirically, this is what happens both for increment threshold and for brightness matching (Hayhoe & Whittle, 1992) when the background is flashed rather than steady, so that the $-x$ term does not have time to develop fully. The role of that term lies in explaining the precise form of isobrightness or t.v.i. curves, not in explaining why the background *brightness* is discounted.

In Equation 7 what corresponds to discounting the background brightness is not the $-x$ term, but the other minus sign in the middle of the equation. The crucial assumption in the model for explaining contrast brightness is that the brightness of a light $(I_b + \Delta I)$ corresponds not to $R(I_b + \Delta I)$, but to the *difference* in response, Δ, between light and surround. If $R(I)$ is a monotonic increasing function of I, then increments and decrements in I yield positive and negative values of Δ. So if brightness in turn is a monotonic increasing function of Δ, any increment must be brighter than any decrement, so the two can never match. Further, because a constant Δ is also the criterion for threshold, the t.v.i. curve is an isobrightness curve. This pattern of results is "complete discounting of the background brightness."

This treatment is somewhat cavalier. As it stands, the model of Hayhoe et al. (1987) does not accommodate decrements, and there is more than one way in which it could be modified to do so. But however this is done, the basic argument will probably stand: What accommodates contrast brightness is the assumption that what we see corresponds not to R but to differences in R. This assumption also accords with the several lines of evidence that suggest the edge-dependence of brightness.

Hayhoe, Levin, and Koshel (1992) extended and refined the measurements of Hayhoe et al. (1987) and found the time course of the $-x$ term to be even slower than before. Little happened in the first 200ms, and the process was not complete for several seconds. They argued that because this is two orders of magnitude slower than *spatial* antagonisms in the retina (center-surround antagonism in receptive fields), it probably represents a slow-acting *temporal* filter. But they argued further, on the basis of Battersby and Wagman's (1964) finding that increment threshold on a flashed background is much less when the flash is smaller than the background than when they are the same size, that there must also be an essentially instantaneous "spatial . . . filter responsible for subtractive adaptation." Another way to put this argument is to say that Battersby and Wagman's discrimination was made on the basis of detecting an edge, and that this could be done instantaneously, that is, when test patch and background were flashed together. It was not necessary to wait a while for the edge to appear. Similarly, the brightness judgments of Hayhoe and Whittle (1992) were based on seeing edges that were instantaneously visible.

This discussion has provided us with some ingredients for an explanation for contrast brightness. Before completing the sketch, we need to consider arguments that the relevant processes are not all retinal.

9.3. A Role for the Brain?

9.3.1. *The Filling-in Problem.*
The idea of contrast coding at edges makes problematic the simple fact that we see uniform regions as uniform rather than as outlines. This has often been thought to call for some filling-in process or rule for extrapolation. By itself, for stimuli in uniform surrounds, the argument depends on an unjustified assumption of isomorphism between percept and neural substrate. But we have only to consider the case of a uniform region on a nonuniform background to see that there is something to be explained here. How do the contrast signals at different parts of the boundary combine to determine the color seen? And various authors at least since Walls (1954) have pointed out that there are several other phenomena, both normal and pathological, that seem to require such a process. One of the most dramatic is the visible filling-in by the surround color of the region from which a stabilized image has disappeared, and beyond if the configuration of contours permits it (Gerrits & Vendrik, 1970;

Krauskopf, 1963). In the binocular rivalry of congruent figures larger than a few degrees, most observers see a curtain of one of the monocular colors sweep across the figure when sufficient of its contour is dominant. Because the neural basis of the alternations is likely to be at the site of the edge signals, this curtain may be filling-in in action. Arend and Goldstein (1987a) argued persuasively that the explanation of various gradient illusions requires the postulation of an active filling-in process. It is also a commonplace idea in computational approaches (e.g., Barrow & Tenenbaum, 1986, p. 38–46). We have no idea what the neural basis of filling-in might be, but it is likely to be beyond the primary visual cortex because units there are unresponsive to uniform areas, and if the rivalry observations are pertinent it must be beyond the site of binocular combination.

9.3.2. Are Contrast Expressions Entirely Retinal?

Filling-in is not the only possible role for the brain in producing contrast brightness. Shevell (1986) argued that isobrightness curves like those in Fig. 2.8 do not represent only retinal function. He generated similar curves in a novel way and found a degree of interocular interaction that showed that they were not monocularly determined. However, Shevell's curves were not isobrightness curves. They were derived by the following ingenious method. A haploscopic display was used like Fig. 2.1(a) but with an additional test patch in one eye congruent with that in the other, so that subjects still saw two test patches but one was monocular and the other a binocular mixture. Isoresponse curves were obtained that showed the value of ΔL required in one component of the binocular mixture (on various background luminances), the other component being held constant, to produce a constant fused brightness. These curves turned out to depend on the value of the constant contribution made by the other eye. Shevell argued that this should not be so if each eye transmitted a single brightness signal to the brain. Such a signal should be represented in the isobrightness curves of Fig. 2.8 and should also be the contribution of each eye to a binocular mixture.

Although this does not show that isobrightness curves for monocular lights are not determined monocularly, it is a salutary reminder that the message from eye to brain is likely to contain more than a single contrast code (section 6.4). It should make us cautious in claiming that the brightness matching results are entirely retinally determined. The brain must be playing some role, or the subject would see nothing, and we have just argued that the brain does the filling in. How much the precise contrast expressions that describe the isobrightness curves in Fig. 2.2 or 2.8 reflect retinal computation, and how much they depend on selection and combination of contrast signals by the brain, remains an open question. I do not yet know of any evidence for a cerebral contribution, but we cannot rule it out. A more positive reason for keeping it in mind is that there are aspects of the contrast brightness data that suggest a perceptual rationale as much as a retinal explanation. For example, the fact that increment and decrement isobrightness functions appear in the contour map of Fig. 2.11 as approximately parallel but with a slope less than 1.0 suggests a possible role in the perception of illumination level. There are more cases like this (equally speculative), and, whereas such perceptually useful features may be incorporated at a retinal level, it also seems possible that they might be introduced in the selection and combination of contrast signals by the brain for particular perceptual purposes.

9.4. A Two-Stage Model for Contrast Brightness

The arguments in the previous three sections suggest a two-stage model. First, contrast brightness depends on a *spatial* difference between local responses, computed by a fast-acting spatial filter operating after a saturating nonlinearity (the R transform). Before the R transform there is a multiplicative gain control that is reflected in the denominators of Equations 1 and 3, and a subtractive term that may affect primarily the exponent n. These processes are probably retinal and could provide one level of explanation for most of the matching, scaling, and discrimination data described earlier.

This leaves unexplained the appearance of the objects as uniform regions, a feature that is not represented in those data. This is likely to depend on a higher level filling-in process. There are several arguments for such a process, and a two-stage scheme could resolve some of the arguments in the surprisingly persistent debate about the neural site of brightness and color contrast effects. Although a retinal locus has been proposed at least since Young (1807), and there is now overwhelming evidence that the neurons of the early stages of the visual pathway are responsive primarily to contrast, it is still being argued that the site of brightness and color contrast may be cortical (e.g., by De Valois et al., 1986, who say "perhaps even a late cortical site"). These arguments frequently hinge on the wide spatial extent and the slowness (section 7.1) of the simultaneous contrast process. Those characteristics certainly suggest cortical processes, but what is cortical may be only the filling in, leaving the basic contrast coding to be done where it is more likely to be done, in the retina; that is, the arguments are right in inferring a role for the cortex, but wrong in assuming that there is only a single neural locus for the interactions shown in simultaneous contrast.

It is tempting to speculate in more functional terms on why luminance is contrast-coded. For example, within the model of Hayhoe et al. (1987) the computation of the spatial difference does not appear as a component of light-adaptation in the same sense that the m and $-x$ terms do, and it would be interesting to consider the significance of this. However, any simple mapping of either component processes or of perceptual phenomena onto function is probably misleading. A full picture of the perceptual and computational advantages of contrast coding would have to consider at least the following: sensitivity to small differences; normalization; reduction of dynamic range; removing additive and multiplicative noise; spatiotemporal funneling of the signal and its relation to eye movements, fovea–periphery duality, and residual nystagmus. Contrast coding, like other aspects of early vision, may have evolved precisely because it had several advantages rather than just one.

10. CONCLUSIONS ON CONTRAST BRIGHTNESS

10.1. An Ambiguity: Perception of Luminance or of Contrast?

This chapter is about brightness, which is apparent luminance. However, some readers may feel a recurrent doubt as to whether it was not really the perception of contrast that was being studied. The physical variables in all the experiments were both luminance and contrast, but the analysis of the data showed that contrast was the main determinant of the responses over much of the range. An ambiguity therefore arises about the psychological variables because judgments explicitly of apparent contrast would often produce similar results. But this close relationship between brightness and apparent contrast is an *empirical* fact about the situations used. In other situations it does not hold. The chapter concerns the behavior in certain situations of a dimension of perceived *color*, and would lose much of its point if subjects had been judging apparent contrast throughout.

However, the question of what aspect subjects were attending to should not be quite so easily dismissed. In the matching and scaling experiments subjects took pains to judge the appearance of the *centers* of the test patches. Matches of apparent contrast in the HSD are possible only in a subregion of the brightness domain, and in that subregion slightly but significantly different results are produced (sections 2.5 and 4). In the discrimination measurements subjects were allowed to use whatever cue worked. At high contrasts brightness was the dominant one, because apparent contrast saturates, but at lower contrasts brightness and apparent contrast were so closely correlated in this experiment, because background luminance was constant, that it was difficult to attend consistently to one rather than the other, and it would probably make little difference if one did so. (Note that contrast discrimination experiments using grating stimuli also allow subjects a variety of cues, and maximum or minimum brightnesses may be used in parts of the range.) The same is true of the scaling experiment, although there subjects were explicitly told to base their judgments on lightness or brightness.

One could say that for contrasts between detection threshold and about 50%, the conditions used in all these experiments (a) make brightness as tapped by all three methods a function primarily of physical contrast but (b) make the psychological variables of brightness and apparent contrast so closely correlated that they are for some purposes interchangeable (when the small differences just cited can be ignored). They form a kind of dual variable for which the name "contrast brightness" may be unintentionally appropriate. We meet a somewhat similar situation in the next chapter with respect to *brightness* and *lightness*, which are normally quite distinct, but in some situations become empirically interchangeable.

10.2. Conclusions

If measured under certain conditions, brightness is closely related to physical contrast. It can be thought of as a bipolar dimension with zero at zero contrast, which I call contrast brightness. Positive values correspond to the brightness of increments and negative ones to the darkness or lightness of decrements, separated by a zero of invisibility. The following features of contrast brightness are revealed more or less clearly by the three techniques of matching, scaling, and measuring discriminability:

1. Separation of increments and decrements. (sections 2.1, 6.8, 7.1)
2. Complete discounting of the background. (sections 2.1, 5.2, 7.2.1, 8.3, 9.2) This is the most puzzling aspect of contrast brightness. It is discussed in section 8 of the next chapter.
3. The Crispening Effect: good discriminability and rapid growth of brightness with luminance, around the surround luminance. Brightness depends on ΔL at low contrasts. (sections 3, 4, 5.2, 6.3, 7.1, 8.3)
4. Good discriminability near black. Brightness depends on L at high contrasts. (sections 2.4.1, 3, 5.2, 6.7)
5. These are all summarized by the approximate invariance of brightness with the contrast expression $W = \Delta L / (L_{min} + L_0)^n$. For cones, $n = 1$ for increments but is slightly >1 for suprathreshold decrements. For rod increments n is slightly <1. (sections 2.2, 2.3, 2.4.2, 6.7, 8.1)

Confirming evidence comes from experiments in which the luminances of patch or surround were modulated in time, and it was shown further that much of the story can be generalized to contrast color. If contrast coding is indeed a very general process, as suggested in sections 1.1, 6.5 and 8.3, then these will be only two further examples among many.

The main thesis of this chapter is that there is a coherent psychophysics of contrast brightness that shows that simultaneous brightness contrast, light adaptation, and contrast coding are closely related. Most of the properties of contrast brightness are in qualitative agreement with current knowledge of retinal function. A role for brain processes was also suggested: to fill-in between edges, to make uniform regions look uniform. A cortical locus for this process is suggested by its wide spatial extent and its relative slowness.

Contrast brightness and its properties should be borne in mind in thinking about any experiment on the perception of brightness and lightness. However, to put it into perspective, we need to consider situations where brightness behaves differently, which is the task of the next chapter.

3 Contrast Brightness and Ordinary Seeing

Paul Whittle
University of Cambridge

1. INTRODUCTION

1.1. Contrast Brightness: A Special Case

In the previous chapter I described three types of measurement that converged on a common picture of "contrast brightness." This was a single bipolar dimension that included the brightness of increments and the darkness (or lightness) of decrements. Contrast brightness was primarily determined by the physical contrast of test light and surround. I argued that it reflected retinal contrast coding. All three types of measurement—matching, scaling, and discriminating—were made in severely restricted situations. One restriction in particular will repeatedly concern us in this chapter: The patches to be matched, scaled, or discriminated either were, or appeared to be, seen against a common background. It was under that condition that brightness was determined by contrast.

In ordinary seeing, local luminance contrast is not the only factor determining either brightness or lightness. For instance, in a natural scene increments are not always brighter than decrements: It is not hard to find weak decrements in a brightly lit region that look brighter than weak increments in a dimly lit part. And there are several studies in the literature showing that lightness can be independent of contrast. This is implied by all demonstrations of lightness constancy in which the immediate surround of the object is held constant in luminance (Henneman, 1935; Hsia,

1943; MacLeod, 1932). In those studies the luminance of the target changed with its illumination, so its contrast changed with respect to the constant surround, yet its lightness remained constant. Gilchrist, Delman, and Jacobsen (1983) provided an even more telling demonstration of the independence of lightness and local contrast. In their experiment a patch of light could be made to appear black, grey, or white without changing either its luminance or that of its background.

1.2. Plan of the Chapter

Nevertheless, contrast brightness is relevant to normal vision. To better understand the relationship between them, I first compare my brightness matching data with some published studies of "brightness contrast" and "lightness constancy" made without constraining viewing to a single background. In some of these experiments the data showed that subjects equated contrast but in others they equated luminance, and in some the results were intermediate. This pattern of results can be understood in terms of different perceptual functions, in particular, two types of lightness constancy: with respect to changing illumination, with reflectances constant, and with respect to changing background, with illumination constant. The evidence for both is considered in some detail.[1]

If the retina codes contrast, then the ability to equate luminance independently of edge-contrast, for which there is good evidence, requires processes analogous to integrating the contrast signal. One possible manifestation of such a process is the phenomenon of *assimilation*, which is the converse of brightness contrast. Some recent work on this is discussed.

The second half of the chapter returns to the discussion of contrast brightness. Distinguishing reflectance from illumination changes is part of the analysis of the scene into objects and lighting and spatial layout that is fundamental to most vision. I argue that this analysis is the key to understanding which situations yield contrast brightness and so can be explained largely in terms of peripheral processes, and which do not and so can be understood only in terms of higher levels.

Although contrast brightness has a relatively coherent psychophysics, some aspects of its nature and phenomenology remain quite puzzling. In particular, its relativity implies a surprising intransitivity of brightness comparisons.

In conclusion, I argue that these chapters point to a revised partition between sensory and perceptual processes in color vision. They also try to

---

[1]Readers who want to avoid experimental detail could skip sections 2.1–2.3, 4.2–4.3.

reconcile two traditions of work on lightness and brightness: one that has emphasized the effects of edge-contrast, and one that has minimized them.

1.3. Lightness and Brightness

This distinction is a recurrent source of confusion in work on achromatic color. In discussing contrast brightness, *lightness* and *brightness* seemed the appropriate terms for the positive and negative parts of the dimension. Their domains did not overlap so no confusion arose. However, in the work discussed in this chapter, their domains do overlap. They can be aspects of the same stimulus; subjects can match one or the other and it matters which they do. We should therefore try to get the distinction clear before going any further.

Lightness means apparent reflectance. *Brightness* means apparent luminance. To obtain lightness matches, Arend and Goldstein (1987b) asked their subjects to make the two test patches look as if they were "cut from the same piece of paper." To obtain brightness matches one asks subjects to make the patches look "like windows into the same room" or "as though they are sending equal amounts of light to the eye." The distinction is not a matter of splitting hairs. It is between two aspects of objects that are as different as, say, color and texture. Judgments of the relative lightness and brightness of the same objects can differ by orders of magnitude. White paper in sunlight and in shadow both look as if they are "cut from the same piece of paper," but we see that the former is sending enormously more light to our eyes. In vision science, because of our predilection for abstraction and reification, we say it has greater brightness, although ordinarily it is more natural to say that it is more brightly lit. But note that even though brightness is the main cue to illumination, the distinction does not correspond simply to the ecologically crucial one between object and illumination, because brightness depends on both reflectance and lighting. Lightness and brightness are therefore two aspects or "dimensions" of any reflecting surface, an obvious fact that has nevertheless generated a remarkable amount of controversy (Lie, 1969), presumably because of the wish for a single dimension of "sensation" to correspond to light intensity. A second distinction arises from the fact that although both terms can apply to surfaces, only brightness can be ascribed to light sources or to self-luminous objects (which is where it is colloquially most natural). For such objects the reflected component is masked by the emitted and so cannot be seen. The existence of the two possible bases for the distinction has led to some confusion in psychological usage as to whether it applies to two aspects of the same object or two classes of object: reflectance versus luminance, or reflecting surface versus light source. *Apparent reflectance versus apparent luminance* has proved a workable distinction that subsumes both.[2]

1.4. Ordinary Seeing?

I should confess that we do not get very far toward the "ordinary seeing" of the chapter title.[3] I discuss experiments in which judgments are made of real objects rather than of patterns of light seen through small holes in optical systems, but the "real objects" are most often two discs, one shaded, the other not. Some further slight complexities are considered, but usually safely under control on a computer screen. Clearly, we still have a long way to go. We have no difficulty in seeing a book of a subdued red color through a cloud of bluish tobacco smoke and illuminated by bright slightly yellowish light. We can see this all at once and about the same area (the red book), even though to make explicit judgments of any of the three aspects of the scene—objects, light, or media—we attend to each separately. What is more, the objects, light, and media may all be nonuniform in color, and any or all of them may be moving. However, simple displays have their merits. We find support for Arend's argument (chap. 4) that center-surround displays are not rich enough to generate by themselves an unambiguous parsing into objects and lighting, but this defect presents an

---

[2]The existence of two terms is of long standing. According to the Oxford English Dictionary, *bright* and *light* derive from Indo-European roots *bkrag* and *leuk*, respectively, both meaning "to shine," but "leuk" having the alternative meaning of "to be white."

The distinction must be to some extent culture relative. It is easier in English than in German, where the same word *Helligkeit* is used for both. We define brightness in terms of light intensity, a relatively modern concept. Our current usage of VDUs may be subtly changing the distinction. It is possible that even our "sensations" may change with changing technology, or at least the aspects of experience that we attend to. An auditory example is the word *volume* on our amplifiers, which most of us take to be merely a synonym for *loudness*, but in William James's *Principles of Psychology* you find a lengthy discussion of the literal "voluminousness" of sounds.

The reader should be quite clear that, although the distinction is *psychological*, it is defined in terms of the *objective* world properties that we are seeing. Its psychological nature is clear from the definition of both terms as "apparent," and also from the fact that we can make it equally well within pictures, regardless of the physical medium. That could be pigment (a painting or photograph), projected light (slides), or an array of light sources (VDU). What matters is the objective situations they are perceived to represent.

[3]Our journey from the psychophysics of simple displays toward the perception of natural scenes may call to mind the following well-known story that appears in many guises. Benighted space traveller in impenetrable rain forest on unknown planet, on meeting small person with long ears: "Sir, can you tell me the way to Alderon?" Person with ears: "If to Alderon going were I, not from here start would I."

3. CONTRAST BRIGHTNESS AND ORDINARY SEEING 115

opportunity in allowing the influence of other factors such as task or perceptual context to be more clearly seen. That is one justification for devoting all this attention to the perception of such simple displays or objects. The others are historical: It is the obvious halfway house between the psychophysics of the last chapter and the real world, and because there is all this literature, we should see what we can learn from it.

2. COMPARISONS WITH OTHER STUDIES

2.1. Candidates for Comparison

There are several experiments suitable for quantitative comparison in that their subjects also matched the centers of two center-surround displays. Some were conceived as measurements of simultaneous brightness contrast, some of lightness constancy. The distinction between these is not clear-cut. It is a matter of their presuppositions, which are often not explicit. The former were intended to measure the effect on brightness, usually thought of as a "sensation," of variations in surround luminance. Brightness judgments, in the work I discuss, can be thought of as comparative judgments of the *apparent amount of light* coming from two regions in the field. Lightness constancy experiments, on the other hand, were intended to measure the effect of different illuminations on lightness, usually thought of as position in the black–gray–white series of surface colors, or *apparent reflectance*. Most of the older experiments, of both kinds, were done with no apparent awareness that in most situations subjects can choose to match either lightness or brightness, with different outcomes (Henneman, 1935, is an exception). This has been demonstrated by several recent studies (Arend & Spehar, 1993a, 1993b; Arend & Goldstein, 1987b; Jacobsen & Gilchrist, 1988a). Of course, there are many differences between the experiments being compared. Differences in the area and duration of the stimuli, over the ranges used, are probably unimportant, as evidence in the previous chapter suggested, and some of the comparisons confirm. In all these experiments the displays are side by side in the field of view, rather than haploscopically superimposed as in the experiments in the previous chapter, so we expect that to affect the results. The point is to see how.

In describing the outcomes of these experiments, I refer to *equating luminance* and *equating contrast*. The reader needs to be quite clear that I use these as descriptions of the mathematical form of the data, not of what subjects were trying to do. Equating luminance means setting a match in which the values of L were equal. Equating contrast means setting one in which the values of contrast expressions such as $\Delta L/L_b$, $\Delta L/L_{min}$, or

Michelson contrast were equal.[4] Subjects in these experiments were asked to match various appearances: brightness or lightness or subjective contrast, but that is not what the terms *equating luminance* and *equating contrast* refer to. They are simply descriptions of the data.

To make precise comparisons, I transformed all the data into trolands and superimposed them in Fig. 3.1 to 3.4 on the isobrightness contours from Fig. 2.11 of the previous chapter.[5] The axes are $\text{Log}L$ and $\text{Log}L_b$, because those are the easiest to interpret for present purposes. Equating luminance corresponds to a horizontal line and equating contrast to one at 45°. The principal diagonal, marked by a dashed line, divides increments from decrements. Isobrightness contours that cross this line therefore contradict the rule that increments never match decrements. Within the shaded region variations of L are imperceptible: All stimuli look black. The coordinates do not represent ΔL explicitly and so make it hard to see what is happening at low contrasts, where the contours bunch together along the diagonal. This disadvantage is unimportant for most of the data considered and is outweighed by other advantages. This has theoretical implications, as we see later.

If I am right that contrast brightness is fundamental to brightness and lightness perception, the transformed data should lie either along these isobrightness contours or deviate from them in intelligible ways.

2.2. Simultaneous Brightness Contrast

Consider first two experiments matching brightness in the presence of different surround luminances. Figure 3.1 shows isobrightness loci from Heinemann's (1955) well-known study of brightness contrast. They are approximately of the same form as my contours. See also Whittle and Challands (1969). Note that Heinemann's decrement curves agree with

---

[4]L is the luminance of the region in question, L_b that of its surround, $\Delta L = |L - L_b|$, and L_{min} the smaller of L and L_b. See Fig. 2.1 of the previous chapter.

[5]I made approximate conversions to trolands, for studies that used the natural pupil, with the aid of the graph in Wyszecki and Stiles (1982, p. 106), assuming that pupil size is determined by the highest luminance in a display. Variations in assumed pupil size make only small changes in the shapes of the curves. Conclusions about, for example, the accuracy of luminance-matching all hold whether the data are plotted in trolands or in cd/m^2.

As described in the previous chapter, the dashed contours in Fig. 3.1 to 3.4 represent detection threshold, and the others brightnesses or darknesses 5, 10, 15, 20 times greater than threshold. For clarity some of the contours near the diagonal are omitted except in the lower left corner.

Fig. 3.1 to 3.4 show only selected data from the various studies. The conclusions I draw, however, are based on considering all of it.

3. CONTRAST BRIGHTNESS AND ORDINARY SEEING   117

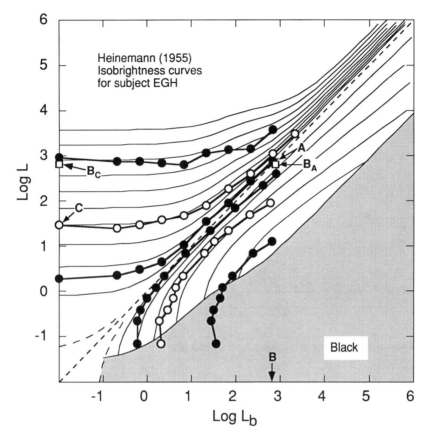

FIG. 3.1. Data from Heinemann (1955). Brightness-matching. Steady lights, haploscopic, fixating. 28' diameter fields centered 52' from fixation point, in 55' surrounds. In Fig. 3.1 to 3.4 the isobrightness contours are from the previous chapter. The letters A, B, etc. are referred to in Fig. 3.10.

mine in requiring larger contrasts at both low and high background luminances than at intermediate ones.[6]

However, not all brightness contrast data agree so well. Heinemann's other subject produced the same pattern of curves but did sometimes match increments to decrements. This was true for both subjects in a later study by Heinemann (1961). Figure 3.2 shows data from one of Jacobsen and Gilchrist's (1988b) variations on Hess and Pretori's (1894) experiment.

---

[6]For the part of the lowest curve that falls well within the black region the value of L is probably irrelevant, as it is for similar points in my data in Fig. 2.8 of the previous chapter.

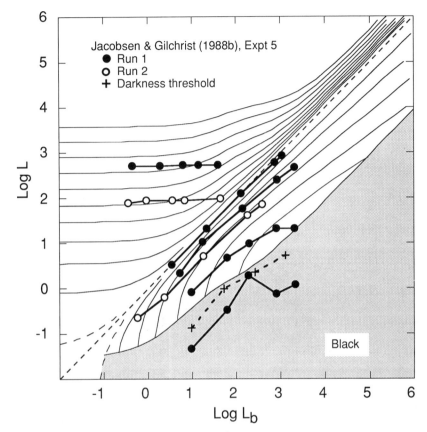

FIG. 3.2. Brightness matching (increments) and lightness matching (decrements) from Jacobsen and Gilchrist (1988b). 1° squares centered in 11° surrounds 65° apart. Steady lights, haploscopic, free viewing.

Their increment curves are flat: luminance is equated. For decrements the agreement with my contours is better. Their "darkness threshold" (crosses) and my "black limit" (the boundary of the stippled region), which are both estimates of the minimum amount of light detectable in a high-contrast decrement, are quite close. They argue that their data within the black region are essentially random.

The isodarkness contour from Jacobsen and Gilchrist in Fig. 3.2 that extends to the lowest values of L_b (unfilled symbols) does not bend round in the way mine or Heinemann's do (this was true also of other curves that are not shown). The likely reason is that Jacobsen and Gilchrist's subjects would have been using rods as well as cones, because there was time for adaptation to the low luminances, and fixation was not restricted. Whereas the other two studies were restricted to cone vision, by requiring fixation and using test patches close enough to the fovea to stimulate few if any rods. In Fig. 2.4 of the previous chapter the scotopic isobrightness curves, approaching the mesopic region from below, continue up into it with remarkably little change of slope, in a similar way to Jacobsen and Gilchrist's isolightness curves, entering the mesopic region from above.

2.3. Lightness Constancy

Next, consider two similar experiments that were intended to study the perception of surface colors under different illuminations. Helson (1943) reported suitable data from a traditional constancy experiment (by Bornemeier after Katz). Subjects saw two rotating discs, made from black and white sectors. One disc and its background was brightly illuminated, and the other in shadow. Subjects adjusted the reflectance of the brightly lit disc to match the other in lightness.[7] The pairs of points joined by lines in Fig. 3.3 show matched pairs. The filled symbols represent conditions in which the backgrounds had the same reflectance. These pairs lie quite precisely on the contours except where they extend to low luminances, probably again because mesopic vision was used. This is classical lightness constancy: Equal lightness corresponds to equal contrast when illumination is the only variable (section 3.2).

A very different result is represented by the outline symbols. Now the backgrounds were black and white rather than being of equal reflectance. These data are totally at variance with the contours (and with Wallach's widely accepted "ratio rule" published 5 years later). The simple fact that two grays can be found that will look the same lightness when one is on white and one on black shows that local contrast is not the dominant factor, because the one on white must be a decrement and the one on black an increment. Most important is the implication of the negative slope of the three pairs marked "A." The upper point in each represents a disc on a black background under bright light, the lower one a disc on a white background in shadow. Luminance is reflectance times illumination, and here the difference in background reflectance was large enough to make the luminance difference *opposite in direction* to the illumination difference. So in this case, far from local contrast mediating lightness constancy, it would actually have counteracted it. Nevertheless, the vertical separation of the points represents a strong tendency toward constancy. If constancy

---

[7]It is not clear whether the experimenter was aware of the possibility of choosing between brightness and lightness. However, in the discussion Helson used the phrase "match in lightness."

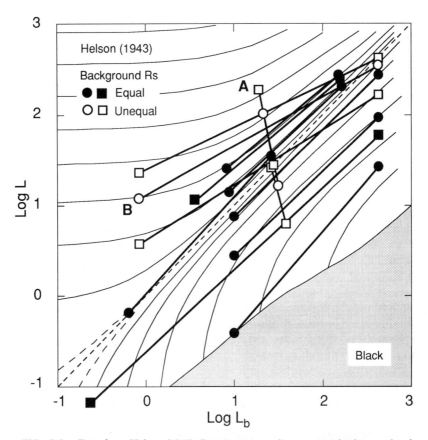

FIG. 3.3. Data from Helson (1943). Rotating sector discs against backgrounds of either the same reflectance (filled symbols) or not (unfilled). One disc and background brightly lit, one shadowed. Subjects adjusted the brightly lit one to match the other in lightness. Symbols of the same shape joined by lines are matched pairs (different shapes only to make the pairing easier to see). Binocular, free continuous viewing. No dimensions given. The contour map here is an enlargement of a portion of that in Fig. 3.1 and 3.2.

were complete the vertical separation should be the same as the illumination difference, which was 1.26 log units. In fact, for A it was 0.74 on average: 60% constancy. Therefore, factors other than local luminance contrast can be very important in lightness judgments. However, local contrast is not irrelevant even here. This can be shown by comparing the A pairs with those marked "B." The backgrounds for B were also black and white, but the illumination difference was the other way round than for A. The constancy for B, aided by local contrast, was 120%, compared to 60% when opposed by it.

3. CONTRAST BRIGHTNESS AND ORDINARY SEEING 121

My fourth and last comparison is with a study by Arend and Spehar (1993a, 1993b) that combined lightness constancy and brightness contrast in one experiment. They used two center-surround displays presented on a VDU (Fig. 4.7 of Arend, chap. 4). Each surround was in turn surrounded by a "Mondrian" border: a region composed of overlapping rectangles of miscellaneous sizes and lightnesses. The luminances were chosen so that both displays and borders looked like matte papers, but with an illumination difference between right- and left-hand displays. The point of the Mondrian borders is to provide a large enough sample of lightnesses under each illumination to produce an unambiguous perception of an illumination difference.[8]

One display was held constant and the simulated illumination on the other varied from trial to trial. Subjects had to adjust the center patch in that display to match the constant one. Figure 3.4 shows 12 iso-appearance curves for one subject. The five points on each curve correspond to five different illuminations on the variable display. The conditions differed in three ways: increment or decrement standard (the large crosses); same or different surround reflectances (filled or unfilled symbols); and three different matching instructions (making 2 × 2 × 3 = 12 conditions). The most unusual feature is the use of three different instructions: (a) Match lightness: Make the patch "look as if it were cut from the same piece of paper" (square symbols); (b) match brightness: "Have the same brightness as . . ." (circles); (c) match brightness contrast: "Have the same brightness contrast with respect to its surround" (diamonds). (I call this *apparent contrast* to avoid confusion with the phenomenon of simultaneous brightness contrast.)

Only one curve crosses the dashed diagonal; that is, the contrast of the match was generally the same sign, increment or decrement, as the standard. Consider first the filled symbols, corresponding to equal-reflectance surrounds. The lightness and apparent contrast matches (squares and diamonds) are almost identical and lie along the contours (like Helson's data, where comparable). Contrast is equated. The brightness matches (circles) have a lower slope, about halfway between equating contrast and equating luminance. This tendency is sufficiently strong that increments are matched to the decrement standard at low illuminations.[9] Brightness

---

[8]Experimenters using real surfaces and illuminations have achieved this by providing various cues to the lighting arrangements. Arend and Spehar's technique offers a route to studying exactly what cues are necessary.

[9]The lightness and brightness data replicate an earlier study by Arend and Goldstein (1987b). Their increment brightness matches were slightly closer to equating luminance. The greater frequency of crossings in Helson's results is just due to using different illuminations and reflectances.

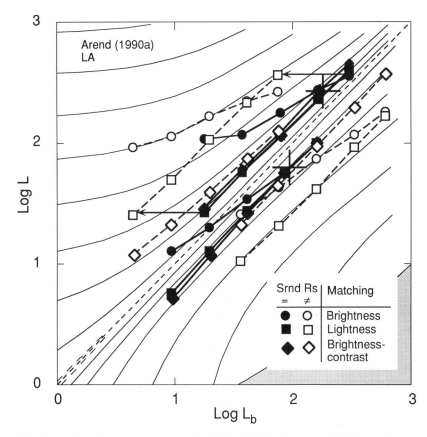

FIG. 3.4. Data from Arend and Spehar (1993a, 1993b). Subjects explicitly matched reflectance, brightness, or brightness contrast on different occasions. VDU display simulating matte paper under different illuminations. 1° squares in square surrounds of side 3°, those in turn surrounded by a "Mondrian" border, 1° wide, composed of rectangles of various shapes and lightnesses. The two displays were 2.5° apart. Steady lights, binocular, free viewing. The contour map here is an enlargement of a portion of that in Fig. 3.1 and 3.2. The large crosses mark the two standard stimuli.

matches would be expected to tend toward equating luminance because luminance, the amount of light coming from the test patches, is what the subject is trying to judge.[10]

The outline symbols represent conditions that were identical except that the surround reflectance of the variable square was reduced by a factor of four for increments and increased by the same factor for decrements. This

---

[10]Even in this experiment, in which unusual care is taken to make the instructions precise, there is still a residual ambiguity. Subjects are asked to make *perceptual* judgments. They are not asked to make the best possible estimate of the relative physical luminances. If they were

corresponds in Fig. 3.4 to the log luminances on the abscissa being 0.6 less for the outline than for the filled symbols for increments, and 0.6 more for decrements. Now consider what happens to the ordinate, the matching luminance the subject sets. The apparent contrast data (diamonds) are the easiest to understand. These lie almost on the same line as before, just displaced diagonally because of the new surround luminances. Contrast is equated, and the match depends only on the luminance of the background regardless of what combination of reflectance and illumination produces it.

The lightness data (squares) are quite different. For increments they are simply displaced sideways, as the arrows show. This demonstrates perfect lightness constancy *with respect to the change in surround reflectance*. When the illuminations were the same, as they were for corresponding pairs of points like those connected by the arrows, the subject set the same luminance to achieve the same lightness. The change in surround reflectance had no effect even though it radically changed the local luminance contrast. We can draw the same conclusion as in Helson's experiment: Luminance contrast can be overridden by other factors as a determinant of lightness. We could conclude the same just from the fact that this isolightness line does not pass through the cross representing the standard stimulus. This implies that L and L_b are not the only factors determining lightness. The decrement data show this too, although for them the displacement from the filled points is not purely sideways. There is also a smaller upward shift in the direction of a simultaneous contrast effect of surround lightness.

This may be clearer in Fig. 3.5. Here the outline symbols have been shifted 0.6 log units to the right for the increment data, and the same to the left for decrements, to compensate for the different surround reflectances. This means that the abscissae are now effectively the *illumination* on the variable display rather than luminance. For increments, this made the lightness data (squares) coincide. To do the same for decrements, a vertical shift was needed as well.

The shifts that made the lightness data coincide in Fig. 3.5 have almost done the same for the brightness data (circles). For Arend and Spehar's other subject this was more precisely so. Therefore, for increments bright-

---

asked to do that, and given feedback, they would after some experience make conscious correction for their perceptual errors, as professional photographers try to do. Then subjects might always equate luminance. But with the present instructions we expect only a tendency in that direction. The residual ambiguity would become apparent if you had professional photographers as subjects. Would they be able to separate their perceptions from their skill, or would their skill have altered their perceptions? My guess is that they would not be sure what to do. Being a subject in this kind of psychological experiment often requires a certain degree of play acting. In this case you have to remember not to act like a professional photographer.

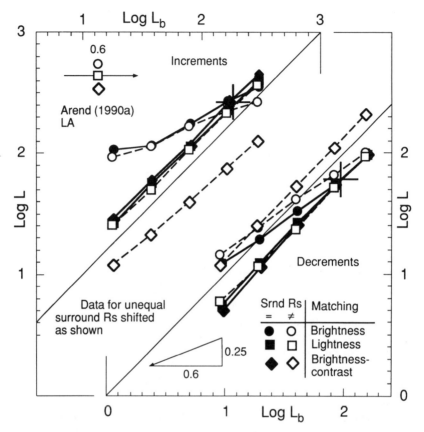

FIG. 3.5. The same data as Fig. 3.4, but with the outline points (unequal surround reflectances) shifted en bloc as shown: 0.6 to the right for increments, the same to the left for decrements with the addition of a vertical shift. See text.

ness judgments are independent of surround reflectance and for decrements there is a contrast effect, just as for lightness. Finally, in Fig. 3.5, note that the shifts that brought the lightness and brightness data together have separated out the apparent contrast data, emphasizing that those matches were determined by luminance, not illumination.

Jacobsen and Gilchrist's results agree with a subset of Arend and Spehar's, with one instructive complication. For increments the agreement is with the latter's brightness matches (both tend toward equating luminance), but for decrements it is with their lightness matches (equating contrast). This makes sense in terms of the instructions that Jacobsen and Gilchrist used. Although they were repeating an experiment on *simultanen Helligkeitscontrastes*, their instructions specified brightness matching only for increments ("where the target appears luminous"). For decrements they in effect asked for lightness matching ("match the shade of gray").

Subjects can also show some tendency toward equating luminance when asked to match the brightness of decrements. We see it in Arend and Spehar's data and it is implied by the reflectance matches in Gilchrist (1988; constancy condition of Expt. 1). Thus these three studies together show that contrast is equated under lightness instructions when background reflectance is constant but there is a tendency toward equating luminance under lightness instructions when surround luminance varies, or under brightness instructions, and both tendencies can occur for both increments and decrements.

2.4. Summary of the Comparisons

We can summarize these comparisons as follows:

> *Brightness* matches sometimes agree with the contours, equating contrast (Heinemann's data in Fig. 3.1), but others deviate toward equating luminance (other data of Heinemann; Jacobsen & Gilchrist for increments; Arend & Spehar).
>
> *Lightness* matches between patches on *backgrounds of the same reflectance* but differently illuminated agree well with the contours (Arend & Goldstein; Arend & Spehar; Helson; Jacobsen & Gilchrist).
>
> *Lightness* matches when the *backgrounds have different reflectances* depend on other factors as well as luminance contrast. When illumination is constant, luminance may be equated. (Arend & Spehar's increments, and implied by Gilchrist et al., 1983).[11]

Note that we are accumulating a list of features of brightness and lightness matching experiments that need to be taken into account in interpreting their results. It includes at least the following:

1. Is the task to match brightness or lightness (or apparent contrast)?
2. Are the test regions increments or decrements?
3. Are they seen as surface or aperture colors?
4. Do the backgrounds appear to have the same reflectance?
5. ... the same illumination?
6. Over what range does background luminance vary? (See following.)

The experiment of matching the centers of two center-surround displays, which at first sight looks so straightforward, turns out to be quite complicated.

---

[11]Not discussed earlier because they were not comparing two center/surround displays.

3. TWO PATTERNS OF RESULTS AND TWO TYPES OF LIGHTNESS CONSTANCY

3.1. Differentiation and Integration

To understand this pattern of results, we need to consider both the purposes and the mechanisms of perception. Whittle and Challands (1969) proposed the following scheme. We suggested that, whereas the retina differentiates and attenuates, the brain can integrate. A subject matching brightness in an orthodox side by side display like Heinemann's can take up different perceptual "attitudes," which correspond on the one hand to different perceptual tasks, and on the other, in terms of brain process, to integrating or not.[12] The retinal signal alone without subsequent integration could be the basis for brightness contrast and for lightness *constancy with respect to illumination changes*, with constant background reflectance. But the integrative brain process, transforming the contrast-coded signal into something closer to a representation of luminance, could be the basis for a second kind of constancy, this time *constancy with respect to surround changes*, with the object under constant illumination, as, for example, when an object moves in front of a variegated background.

To clarify the argument let me first make the passage from data to theory more explicit. The data from these various experiments show two recurring patterns: equating contrast[13] and equating luminance. These patterns suggest both mechanisms and also functional interpretations in terms of the real-world perceptual tasks corresponding to each. The mechanisms suggested by equating contrast, those of adaptation and contrast coding, were discussed in the previous chapter. The integrative mechanisms suggested by equating luminance are discussed later.

In the next two sections I say a little more about the possible perceptual functions. The two types of lightness constancy are not the only ones. The perception of several different properties of the world depend on judgments of light intensity, not only surface lightness and the brightness of

---

[12]*Tasks* is not quite the right word because of its connotation of conscious purpose, which is only appropriate in some situations. More often we have a mode of operation triggered by the stimulus situation. We can talk in terms of *goal, purpose,* or *function* on the one hand or in terms of stimulus *information* or *cues* on the other. This is the familiar difficulty in psychology of finding terms that are at home on both sides of the Cartesian division of mind–body and voluntary–involuntary. In the present context we are concerned with the *judgments* that we are trying to make, or for which our visual system is spontaneously providing us with evidence. Battles are fought over this distinction, but I do not join them here.

[13]I defined equating contrast as equating contrast expressions. I do not discuss any further the departures from Weber's Law at lower luminances. This is likely to occur if luminance is made low enough, though the preceding discussion suggested that Weber's Law is maintained to lower luminances if vision is not restricted to cones.

light sources but the depth of shadows, the opacity of media, and the intensity of illumination. This is to include only those that are explicitly intensity judgments. There are others like shape-from-shading that depend tacitly on judging intensity. Furthermore, each can be perceived in such different contexts that a two fold diversity is implied. This diversity must be borne in mind when considering the perception of brightness and lightness. It has too often been unimaginatively assumed that only one or two perceptual functions are served by judgments of light intensity. One reason this is important to visual science is that all these judgments can be made automatically, so that if we ask subjects for "brightness judgments," for example, without further clarification, or without thinking what intensive dimensions can be perceived in the display, we may well trigger mechanisms or strategies that are appropriate to surfaces or shadows or filters or highlights.

3.2. Equating Contrast

It has often been pointed out that the contrast between reflecting surfaces is independent of overall illumination, because it depends on only the ratio of reflectances. Therefore if lightness is a function of only contrast, it too will be independent of illumination. This is what has traditionally been meant by *lightness constancy*. Equating contrast when matching lightness is experimental evidence for it.

However, equating contrast may also reflect a quite different visual function of obvious survival value. This is seeing illumination patterns: shadows, shading, light patches, and highlights. A current vivid example is the artificial situation of seeing projected slides. A natural example is seeing the shadow of a predator. If lightness constancy were perfect in the sense that illumination gradients were completely discarded, these would be invisible, because they are only patterns of illumination. But of course we see them. This is not surprising if we think of vision as primarily seeing contrast, but it is an instance of a remarkable achievement of the visual system: keeping different forms of information (or levels of processing) available for alternative interpretations.

Our perception of the colors of objects in slides is to a considerable extent independent of the color and lightness of the screen, even when that can be seen if we attend to it (Cramer, 1923). Physically, this is illumination constancy in the presence of reflectance changes. It is the converse of classical lightness constancy, though it has the same mathematical form: invariance with respect to a uniform multiplicative component. Bergström's idea of the removable common component (chap. 6) neatly encompasses both. However, I know of no experiments on matching illumination patterns that the subjects clearly perceive as such. My guess is that contrast would be equated under some circumstances and luminance under others, but we do not yet know.

3.3. Equating Luminance

Woodworth (1938), in line with many other authors, wrote that judgments in constancy experiments "usually lie between two extremes, one conforming to the stimulus and the other conforming to the object" (p. 605). In our case this means conforming to luminance or to reflectance. However, equating luminance can be interpreted not as conforming to the stimulus but as another kind of object judgment.

Equating luminance corresponds to all situations under which we see an object of constant luminance as having relatively constant lightness or brightness in spite of variations in the surround luminance. The surround could be foreground or background and it could vary in reflectance or illumination. The defining feature of these situations is that, although the luminance of the surround varies, the illumination on the object is perceived to be constant. The second kind of lightness constancy suggested earlier is the chief example of this. This happens whenever one picks something up and carries it around, so that it is seen against different pieces of furniture or floor. Another instance is seeing an object behind or between other nearer objects. The nearer objects are sometimes a mesh or network, such as is formed by trees or other vegetation. Another situation, which might better be called brightness than lightness constancy, is a light source, or a window into another scene, with a surround changing in luminance or color. Veridical perception in any of these situations requires discounting the *surround*, whether background or foreground; that is, perceiving luminance, not contrast.

In 1969 we suggested the existence of lightness constancy with respect to surround changes for both empirical and theoretical reasons. Empirically, it seemed to us from informal observations that it existed. An object moved over different backgrounds does not seem to change much in lightness. To get a good look at a sample of cloth, you may pick it up and take it to a good light, but you do not worry about what background is behind it. It is as though the background does not matter. It is not that there is no simultaneous contrast effect, if you look for it; just that it is amazingly small. If brightness was always contrast brightness, such objects would flash on and off all the time as they changed from being increments to decrements. They do not.

The theoretical reason was that we need to classify this as a constancy but now vision is thought of as seeing contrast. When it was thought of as seeing light, constancy with respect to surround changes did not seem to require explanation, because such changes do not affect the light in the focal region itself. But now we know that the retina codes contrast, and in the light of measurements like mine that show the enormous quantitative extent of brightness contrast, it becomes a puzzle how the constant appearance of an object is maintained against different backgrounds. This becomes a form of color constancy just as remarkable and in need of explanation as constancy with respect to illumination. This idea seems to be gaining some currency, and not only among the authors of this book (see the chapters by Gilchrist and Arend as well as my discussions of their work). For example, Brainard and Wandell (1986) wrote: "In our view, color constancy should be defined as the maintenance of color appearance despite variations in the *color of nearby objects* and despite variation in the spectral power distribution of the ambient light" (italics added, p. 1651). And Brenner, Cornelissen, and Nuboer (1987) wrote, "concentrating exclusively on the contrast between an object and its direct surrounding implies that moving the object in front of a different background must change its colour. Although this does happen, the effect is extremely weak compared with what one would expect if only local edges contributed to the perceived colour" (p. B9).

3.4. Experimental Evidence for the Two Kinds of Constancy

The distinction between the two types of constancy has been given more empirical substance by the work of Gilchrist and his colleagues. Gilchrist et al. (1983) manipulated the context in such a way that the black–white background of a traditional side by side simultaneous contrast display— two gray patches of equal luminance on black and white backgrounds— was perceived to be either two different illuminations falling on background and patches, or two different reflectances behind the patches with everything under one illumination. Lightness judgments of the center fields followed contrast or luminance, respectively. These correspond to the two types of match and the two types of constancy, occurring in their appropriate conditions, that Challands and I had postulated. As Gilchrist (1988) pointed out, Evans (1948; see quotation in section 3.6 following) had made the same observation, and Woodworth (1938) is only one among many older texts in stating that contrast is "only slightly in evidence when surface colour is clearly perceived" (p. 619).[14]

Further evidence for constancy with respect to changing surround

---

[14]Woodworth's (1938) excellent discussion anticipates many things in this chapter. The full quotation reads: "With regard to contrast, if it is peripheral or at least sub-perceptual in origin, the fact that it is only slightly in evidence when surface colour is clearly perceived means that in some instances object colour is perceived *in opposition to contrast*. Contrast would then amount to an illumination for which correction is made" (p. 619).

reflectance, for both brightness and lightness, is provided by the results of Arend and Spehar in Fig. 3.4. The lateral shift between filled and unfilled symbols shows that subjects set the same luminances to match the same standard lightness or brightness in the presence of two different surround reflectances. Assuming that matching is transitive, this shows that lightness and brightness were independent of surround reflectance.

This experiment provides unusually strong evidence for lightness constancy in general. We usually say that data are evidence for a constancy if they accord with the hypothesis that judgments of some quality are invariant when the corresponding objective quality is invariant. This is just saying that subjects can indeed perceive the objective quality under different circumstances. Denoting the constancy by the subjective term, as in *lightness constancy*, points to the observation that there is usually constancy of appearance rather than just correct judgment. In most cases the inference to a constancy is quite weak. For instance, an isolightness curve may only approximately correspond to constant physical contrast, and the alternative interpretation, that contrast and not reflectance was the object property being judged, is not excluded.

In Arend and Spehar's data, however, the existence of the two types of constancy provides a pattern of evidence converging on the conclusion that lightness is constant if and only if reflectance is. Constant lightness corresponded to constant contrast when only illumination was changed, and to constant luminance when only surround reflectance was changed. The conclusion that encompasses both is that lightness is constant if and only if reflectance is. It is hard to think of an alternative interpretation. Further, the constant contrast lines are almost exactly so, and the lateral shift in the case of increments corresponds quite exactly to constant luminance. The data therefore permit an unusually strong inductive inference to lightness constancy.[15] They not only give evidence for both types of constancy but the occurrence of the two together strengthens the evidence for each of them.

Helson's data in Fig. 3.3 provide similar converging evidence, but the inference is less strong because the fit is less precise. This experiment also demonstrated both types of constancy, though less clearly than Arend and Spehar's because there was no condition in which only surround reflectance was changed. Another well-known study by Helson (1938), in which both the color and intensity of illumination were varied, found marked

---

[15]There seem to be two reasons for the good constancy in Arend and Spehar's experiments. First, they took care to give precise instructions. Second, the Mondrian border facilitated the accurate equation of reflectances across scenes even when surround reflectances and illuminations were both different, by providing a series of grays in each display. The subject's setting put the two patches in the same relative position in the two series.

changes in lightness judgments when background reflectance was changed. This kind of constancy is certainly not invariable, and we as yet know little about the factors that influence it. One factor, the range of surround luminance variations, is discussed later.

Of course, this division into two types of constancy is partly for purposes of exposition. Commonly, both illumination and surround reflectance vary. Also there are other possible variants, such as when illumination, rather than reflectance, varies on the background but not on the object. I know no experiment that has explicitly studied that situation. However, the general conclusion is that we have considerable ability to judge surface color in spite of variations both in lighting and in the color of neighboring objects.

3.5. The Perception of Lighting

How good are we at judging lighting, the aspect of the world complementary to surface color? This question has often been raised but there is still little evidence. *Lighting* includes light sources and illumination. The illumination may be general or local, either of a scene open to us, or one that we see through a window or other aperture; it includes highlights, the depth of shadows, and the darkness of holes. I cannot attempt a review here, but two of the experiments already discussed provide some evidence.

Gilchrist et al. (1983) asked their subjects for judgment of illumination as well as of lightness. They had to select the same illumination from an array on a card in front of them. Their judgments were quite accurate and were indeed the evidence that they were seeing the background as intended, as two different illuminations of one gray or vice versa.

Arend and Spehar's instruction to match brightnesses probably influenced subjects towards seeing the focal regions as light sources or windows or locally increased or reduced illumination. Increments can be seen as light sources. Decrements cannot, but they can be seen as either a window onto a darker scene or a shadow. Apertures or windows onto some other scene are common in natural environments, not just in houses. The brightness judgments lay between equating contrast and equating luminance. For increments, they showed as much independence of surround reflectance as the lightness judgments.

We know little about judgments of the depth of real shadows or of the darkness of holes. I gave some reasons earlier (section 3.2) why the darkness of shadows would depend on contrast, but because holes are, as it were, negative sources, and because the discriminability of dark decrements depends on L rather than ΔL (see chap. 2), it seems possible that their darkness might depend primarily on luminance.

3.6. What Happens in Experiments on Brightness Contrast?

Unlike constancy experiments, studies of brightness contrast usually use displays that are ambiguous as to whether the backgrounds differ in reflectance or in illumination. Indeed, that difference is usually ignored in such experiments, which are both conceived and analyzed purely in terms of luminance relations. The degree of ambiguity depends on several factors, particularly on how much the subject can see of the texture of surfaces and other aspects of the physical setup. Matches in such displays equate either luminance or contrast or something in between, often in the same experiment. This suggests that subjects in this ambiguous situation can take one or the other attitude or vacillate between them (Parrish & Smith, 1967). *Interpretation* would perhaps be a more appropriate word than *attitude*, though it is too intellectualistic a term for what is often just "seeing as," such as seeing a dark patch as a shadow. Whittle and Challands (1969) used the word attitude partly because we had in mind global modes of seeing like Katz's "subjective attitude," and also because we wanted a rather vague term that would be appropriate to subjects' bewilderment confronted with the impoverished stimuli of psychophysical experiments. Such bewilderment was described by Evans (1948) in the quotation that follows and was well documented by Burgh and Grindley (1962). Our idea was that although that poverty is no problem for yes–no threshold judgments, it becomes a problem for judgments like equating the brightness of lights in different surrounds, because the perceptual mechanisms or strategies we adopt for such tasks are probably the same ones that are involved in seeing a dark patch as a shadow. Because the impoverished stimuli are ambiguous with respect to such interpretations, subjects are at a loss and fluctuate between different ways of parsing the display into surface colors and lighting.

We did not suppose that they were usually aware of clear-cut alternatives, like the reversals of a Necker cube; that is, however, what Evans (1948) suggested in describing his replication of Hess and Pretori's experiment:

> One fact, however, stood out rather clearly. As soon as the brightness difference between the two surrounds became very large some sort of attitude *had* to be taken toward the figure in order to obtain consistent data. At large differences the appearance of the central squares took on an indeterminate, fluctuating sort of quality, and the observer was equally dissatisfied with the match over a relatively large range of intensities. Binocular vision was used in a room which was not wholly dark. Under these conditions it was quite apparent that there were four surfaces visible, and it became possible to ask the observer to consider their appearances in either of two

quite different ways. The whole figure could be seen as a cube viewed from one edge with gray squares on each of two sides, or it could be seen as two apertures through which what appeared to be a single separately illuminated surface was seen. The results obtained by asking the observer to take either one of these two attitudes were quite satisfying. The indeterminateness ceased at once. When the perception was that of a cube with gray squares the matches moved far in the direction of the ratios required by complete brightness constancy. When the perception was that of a uniformly illuminated rear surface seen through two apertures, the results moved equally far toward having *identical* intensities (luminances) on the two sides.[16] (p. 166)

In practice experimenters have taken various measures to get better data, usually to reduce the tendency toward equating luminance. Discussing these measures is part of the general question of which situations produce contrast brightness and which do not. Before we can deal adequately with that, we need to consider two other matters.

3.7. Surround Reflectance Versus Illumination, and Lightness Versus Brightness

I distinguished lightness and brightness at the outset. But in discussing sources of ambiguity in experiments, the focus has been more on the question of whether *surrounds* differed in surface color or in illumination. A brief discussion of the relation between these distinctions may be helpful.

Reflecting objects can be compared in either lightness or brightness, and, when we can see an illumination difference between them, it may make a big difference which we are doing. Arend and Spehar's results in Fig. 3.4 and 3.5 confirm this. The importance of seeing whether *surrounds* differ in surface color or illumination or both is just that it is the main clue to the relative illuminations of the *objects*.

The two choices, brightness or lightness, and same illumination or not, are different in nature. They generate different kinds of ambiguity. Ordinarily, we see at once whether two objects are under the same or different

---

[16]See also Gilchrist (1988). Evans also reported that his observers fixated between the squares to judge lightness but looked from one to the other to judge luminance.

When I wrote the discussion to Whittle and Challands (1969), I was under the impression that I was thinking it out for myself. However, while writing this chapter I found some old notes of mine on this passage of Evans. Most scientists will be familiar with this kind of embarrassment. I would like to acknowledge now the many valuable insights contained in Evans' work; far more than I have yet assimilated. He also worked with the HSD, for example, in Burnham, Evans, and Newhall (1957).

illuminations. But alternative interpretations are generally available, even though we are unaware of them. For instance, we can decide to see a scene as a picture and judge what lightnesses would be required to depict it in the (uniformly lit) picture. Or when we are actually looking at a picture, we can attend to either the different depicted illuminations within it or the uniform illumination falling on it. These ambiguities are between different perceptual interpretations or organizations, as with a Necker cube. Whereas the choice between lightness and brightness of surface colors is not a matter of perceptual ambiguity, they are just two aspects that are usually both there to attend to and report on if we wish, just as hue and texture and saturation also are.

Thus our perception of lighting sets the scene, and the choice between lightness and brightness is one of the distinctions we can make within it. When we see no illumination difference, the distinction between lightness and brightness collapses. This happens in the HSD (Haploscopic Superimposed Display, shown in Fig. 2.1a of the previous chapter). We can still try to compare either reflectance or luminance, but they will be perfectly correlated so it will make little difference.

4. INTEGRATION AND ASSIMILATION

4.1. The Need for Integration: Recovering Luminance From Contrast

Contrast brightness is a function of contrast. The evidence in the previous chapter belongs with a large body of other psychophysical and neurophysiological knowledge justifying the claim that vision is AC coupled (e.g., Laming, 1986); that is, the eye responds to spatiotemporal transients and filters out DC or low spatiotemporal frequencies. However, the possibility of equating luminance shows that to some extent DC information can be recovered. Some kind of integration must take place at higher levels of the visual system. The same need was pointed out in the previous chapter in the context of filling-in: that we see uniform areas rather than just the contrast at edges.

Possible integrative processes have been suggested in several contexts (e.g., in Land's retinex theory, or the computational versions of it proposed by Blake, 1985a, 1985b, or by Arend and Goldstein, 1987a, or in the "edge-integration" of Gilchrist et al., 1983). The development of adequate models of these processes is a major theoretical challenge. Thinking of them as only integration to recover luminance may be misleading. As Gilchrist et al. pointed out, classification of edges is as important as integration. What is recovered is not luminance but surface color, illumination, transparency, and, intimately bound up with them and probably transcending them all in importance, spatial layout. In some experiments this process allows subjects to equate luminance, but luminance may not be represented at any level independent of object qualities. (See Arend, chap. 4, for some discussion of the different levels involved.) I do not discuss general models here. The aim of this chapter is the less ambitious one of considering when integration occurs and when it does not.

We have seen that there is evidence for good lightness constancy in the presence of surround variations. But compare this with the enormous influence of local contrast in the phenomena described in the previous chapter, and in countless other aspects of vision. It is as though for most purposes vision uses a contrast metric, but that it is able to change when required to something closer to a luminance metric. When the latter happens, both aspects of the contrast metric disappear: differencing and attenuation. Not only does equating contrast go, but also the manifestations of the coding of ΔL: the Crispening Effect and its associated large improvement in discriminability. In a variant of Arend and Spehar's experiment described earlier, the standard was changed from an increment to a decrement by changing the reflectance of its surround from one Munsell Value step above it to one step below. This had virtually no effect on the match even though it is a change that ought, at first sight, to produce a large change in lightness because of the Crispening Effect. The integrative processes that allow luminance to be equated had apparently compensated for the Crispening Effect also. This is the theoretical reason, alluded to before in justifying the choice of axes for the contour maps in this chapter, why it is not as necessary to represent ΔL explicitly for this kind of experiment as it was for those in the previous chapter.

The range over which the integrating processes can counteract contrast coding is probably quite restricted, perhaps to the range of luminances in a scene under a single illuminant. The contrast coding Weber behavior describes visual function over a range of background luminances of 9 log units or more. In the experiments in which luminance was equated, L_b was varied by no more than 1.5 log units, and although the isolightness contours crossed the increment–decrement divide, they did not go very far across it. If the range had been longer, using haploscopic displays to avoid glare and adaptation effects, but with the surrounds side by side, it is unlikely that subjects would have continued to equate luminance. We know, for instance, that if we hold up a piece of gray paper against the sky it looks black. However, we noted in the previous chapter (section 6.1) that retinal contrast coding is only unequivocally demonstrable over a restricted range of contrasts, so the integrative processes may be needed only within that range. There was evidence that outside that range the retina makes luminance available for some purposes, such as the perception of intense light sources or of black holes.

When luminance is equated, the judgment is still relative to a context, although one wider than the immediate surround. It is not what is discussed under the heading *absolute luminance perception*. It is still not clear how much that exists, but it is unlikely to be sufficiently accurate to mediate the accurate equation of luminances seen in some of the aforementioned experiments.

4.2. *Assimilation:* A Manifestation of Integration?

Shapley and Reid (1985) recently put forward a two-process scheme like that of Whittle and Challands (1969). They argued as we did that whereas brightness contrast depends on only the adaptive effect of a background, a second, integrative, process depends on the subjective brightness of the background. They do not motivate their two processes by the distinction between two kinds of constancy. Indeed, they seem to leave the function of the second process mysterious. However, they identify it with *assimilation*, demonstrate it experimentally in a manner similar to Arend, Buehler, and Lockhead (1971, Fig. 2e), and Gilchrist (1988, Fig. 4D) and measure it, finding it less strong in their situation than brightness contrast.

Assimilation is a phenomenon opposite to brightness contrast: Increasing the brightness of a region *increases* the brightness of a neighboring region. The term was introduced in the context of color appearance by Helson in 1943, to refer to what had previously been called *von Bezold's spreading effect*. Figure 3.6 shows an example. Because assimilation implies

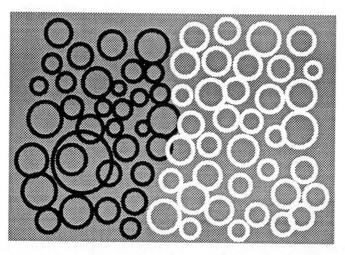

FIG. 3.6. Assimilation: The gray background on the left at first sight looks darker than that on the right. But if you fixate between two neighboring circles, one with a black outline and one with a white, the effect often reverses into the usual brightness contrast effect for the gray regions within the circles.

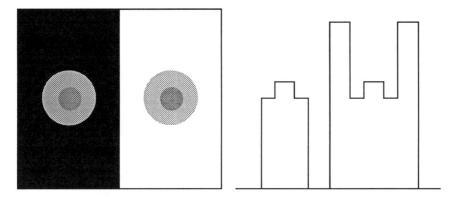

FIG. 3.7. Display used by Shapley and Reid (1985) to study "assimilation." The luminance profile is shown on the right.

that there are processes counteracting the determination of brightness or lightness purely by local contrast, it is of considerable interest in the present context. It could be a manifestation of the hypothetical integrative processes, which also in general counteract the effect of local contrast and thus be a useful tool for studying them.

Two properties of assimilation, in addition to the direction of the effect, suggest links with the second kind of constancy described earlier, that with respect to surround changes. One is its lability. Most people at first see a strong assimilation effect (in Fig. 3.6 the background looks darker on the left), but if you attend to comparing two small areas of background, assimilation usually reverses into brightness contrast (Burnham, 1953; Festinger, Coren, & Rivers, 1970). This is reminiscent of the lability of observers' attitudes in brightness contrast experiments. The second property is that assimilation is particularly strong when the inducing pattern is a network of thin lines (Helson, 1963). Now this is a feature of one common situation in which we need to perceive object color independently of its surround: When we are looking at it through a mesh of vegetation for example. In this situation it could well be helpful to have a mechanism to counter the effects of contrast coding and so yield something closer to the perception of the color of the object when not seen through the mesh. It is not clear why it should overshoot and generate the opposite illusion, although like brightness contrast (Gilchrist, 1988), this could be stronger in artificial situations than in natural scenes.

The type of assimilation studied by Shapley and Reid (1985) is somewhat different. Figure 3.7 shows a typical display. The two center-surround targets have the same luminances. The brightnesses of the surrounds are altered by brightness contrast with respect to the regions outside them and carry the brightnesses of the center patches along with them to some extent. The latter effect is what Shapley and Reid call assimilation. It has the same direction as von Bezold's effect; that is, center and surround brightnesses change in the same, rather than opposite directions. However, it does not seem to show the lability. In my experience, and that of subjects in the experiment by Shevell, Holliday, and Whittle (1992), Shapley and Reid's phenomenon remains completely stable under focal attention. Perhaps there is more than one type of "assimilation."

Burnham (1953) also suggested there might be more than one mechanism counteracting the effects of contrast coding. A tempting explanation for von Bezold's effect is simple averaging in a low spatial frequency channel. The lability could then be an example of rivalry between channels corresponding to different spatial frequencies, as has been reported in other contexts (Campbell & Howell, 1972, on monocular rivalry). Averaging, with appropriate spatial characteristics, is the type of integrative mechanism that most obviously predicts assimilation, whereas, for example, accurate luminance edge integration across a scene would not predict it. However, this is not the place to choose among the great variety of hypothetical integrative mechanisms.

4.3. Is Seeing the Difference Between the Surround Luminances a Necessary Condition for Assimilation?

Contrast is equated most reliably and precisely when subjects cannot see the separate surround luminances, as in the HSD. Conversely, both Whittle and Challands (1969) and Shapley and Reid (1985) argued that seeing them is a necessary condition for assimilation. Because this is a potentially important claim, it is worth examining the arguments for it more carefully.

Whittle and Challands (1969) argued that there must be an additive effect of perceived background brightness on a superimposed patch for three reasons: (a) We measured such an effect (Fig. 5 and Fig. 6 of that paper). (b) It could explain the direction of the discrepancies between our results in the HSD and those of other experiments in which the background brightnesses were seen side by side rather than haploscopically superimposed. The results of such experiments always deviated from ours in the direction to be expected from a positive influence of background brightness when the different brightnesses could be seen; that is, they showed less brightness contrast. (c) By various arguments we eliminated two other possible routes for the background to influence the brightness of superimposed patches (namely, physical addition of light and its adaptive effect). Shapley and Reid (1985) argued that it was the brightness of the surrounds that was important because they varied brightness, and not luminance. Arend et al. (1971) used a different technique to change surround brightness without changing luminance: placing a Craik–Cornsweet

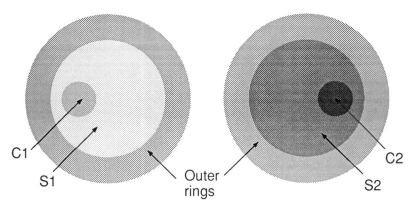

FIG. 3.8. Display used by Shevell et al. (1992) to study "assimilation."

edge in the background between the two patches to be compared. This too produced an assimilation effect.

Shevell, Holliday, and Whittle (1992) showed directly that assimilation occurs in side by side displays but not in the HSD, where the surround brightnesses cannot be seen. A brightness match was set between C1 and C2 (Fig. 3.8), on unequal surround luminances, with the displays seen side by side as shown but without the outer rings. The results showed the expected brightness contrast effect. The outer rings were then introduced, intermediate in luminance between S1 and S2. This *increased* the brightness difference between S1 and S2, by brightness contrast, but *reduced* the luminance difference between C1 and C2 required for a match. This is Shapley's "assimilation."[17] However, when the two displays were superimposed haploscopically, the basic brightness contrast effect of S1 and S2 on C1 and C2 was unaffected, but the assimilation effect of adding the outer rings disappeared. Shevell et al. also point out that their experiment shows that the assimilation effect is mediated by processes occurring after the signals from the two eyes join, because unless the conditions for it are present in the cyclopean field it does not occur. This supports the earlier claims that the effect is cortical.

All these lines of evidence suggest that seeing the separate brightnesses of the surrounds is necessary for assimilation. An alternative interpretation is that it is the edge(s) between the surrounds that must be seen, because the HSD removes such edges as well as masking the brightnesses. Assimilation might depend on integration over the edge(s). Figure 3.9

---

[17]To follow what is happening, consider just the left-hand display and let $L(C1)$ denote the luminance of C1 required to make it some particular constant brightness. S1 darkens C1 by brightness contrast, so $L(C1)$ must be increased to compensate. The outer ring brightens both S1 and, by assimilation, C1, so $L(C1)$ can now be reduced.

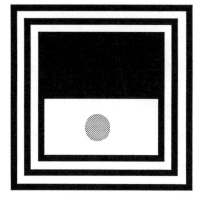 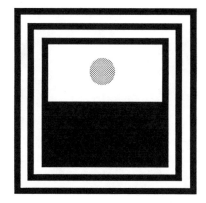

FIG. 3.9. If these displays are binocularly fused, the dividing contour alternates in polarity because of rivalry. If the relative lightness of the gray circles depended on edge integration including the contour currently visible between them, these alternations ought to be accompanied by large changes in the relative lightness of the circles. Are they?

allows an informal test of this idea. If the two displays are binocularly fused, it is possible to see the dividing edge alternating in polarity because of rivalry, whereas the gray circles are relatively undisturbed because each dominates over the corresponding uniform area in the other eye. Now, if the relative lightnesses of the gray circles depended on a process of edge integration over whatever contour is currently visible between them, the alternations ought to be accompanied by large changes in their relative lightnesses. This does not seem to happen, which suggests that assimilation and/or edge integration do not operate in this way.

4.4. Conclusions on Assimilation

The phenomenon of assimilation is opposite to brightness contrast: Increasing the brightness of region A *increases* the brightness of a neighboring region B. There may be two kinds of assimilation. Classical assimilation (von Bezold's spreading effect) is strongest when region A is composed of thin lines. It is a labile phenomenon that readily reverts to its opposite, brightness contrast, if focal attention is given to region B. The assimilation of Shapley and Reid (1985), on the other hand, does not require A to be composed of thin lines and is not labile. This kind of assimilation only occurs when two assimilation-inducing regions can be *seen* to differ in brightness. (This is not required for brightness contrast, where it is the luminance, not the brightness, of the surround that is important. We do not know whether it is required for classical assimilation, though that seems likely.) Both types of assimilation may be manifestations of the integrative processes that lead to equating luminance rather than contrast. The lability, the effectiveness of thin lines, and the requirement of seeing the brightness of region A all support this idea.

5. CONTRAST BRIGHTNESS: WHEN AND WHY?

I now return to contrast brightness, the main topic of these two chapters. In section 3 I argued that the variable results of matching experiments could be understood in terms of the different uses of luminance information in natural scenes. Some would require judging luminance and some contrast, and the evidence so far as it went confirmed that the appropriate match was produced when the corresponding perceptual goal was called forth, either automatically by the stimuli or voluntarily in response to instructions. In this section I take up the question of why contrast brightness is invariably seen in the HSD, and try to give a more general account of when it is seen and when not.

5.1. What Conditions Produce Contrast Brightness?

Experiments in the HSD escape the ambiguity of many studies of brightness contrast by using a display that is even more impoverished. The superposition of the surrounds in the binocular field makes the patches appear to be on the same background, and this seems to lock subjects into equating contrast. That is what the data showed, and the subjective experience was that there was no ambiguity, no possibility of different attitudes.[18] The obvious explanation is that, when the two patches are seen against the same background, only relative (contrast) information is available. Explicit representation of absolute luminance has been lost at an early stage and cannot be recovered in this situation. The HSD abolishes the problem of interpreting the backgrounds as differing in reflectance or illumination by preventing subjects from seeing that they are different at all. An even more radical maneuver is to present the backgrounds as stabilized retinal images, so that not just the difference between them but the backgrounds themselves disappear. Experiments by Yarbus (1967, Fig. 51) and by Piantanida and Gilchrist (see chap. 1) suggest that that has

---

[18]Whittle and Arend (1991) compared the variability of color matches set in the HSD and in a similar display but with surrounds side by side. The conditions were identical except that in the latter case half the surround was blanked off in each eye. In the side by side case standard deviations of repeated matches were on average more than twice as big as in the HSD, and in some side by side conditions the two subjects showed large and consistent differences that were completely absent in the HSD.

the same effect as the HSD on the appearance of superimposed unstabilized increments and decrements.

The data on luminance discrimination and on equal-interval scaling described in the previous chapter showed many of the same features as the matching experiments. It was as though judgments of contrast brightness underlay all of them. In the discrimination and scaling experiments there was only one background: Hence the generalization at the start of this chapter, that contrast brightness is the rule when the lights to be compared either are or appear to be on the same background. Work on contrast color provides additional examples (section 7.2 of previous chapter). For instance, the results of Helson (1938) and Walraven (1976, 1981) seemed to follow contrast colors, and their subjects were looking at objects on a single background and comparing them with remembered hues.

However, judgments sometimes follow contrast brightness even when you can see separate backgrounds.[19] Manipulations less radical than the HSD or stabilization produce contrast brightness. Heinemann used very small, probably textureless displays, presented haploscopically so that each eye could adapt separately. Wallach used a 20° separation between his targets (Jacobson & Gilchrist, 1988b), so that subjects would have to look from one to the other to compare them. In the chromatic experiments of Whittle and Arend (1991), merely flashing the test patches, on steady, textureless, nonoverlapping backgrounds, produced much the same results as the HSD. In the experiments of Magnussen and Glad and colleagues (section 7.1 of previous chapter), the patch or surround luminance was modulated in time. More suggestions can be found in the older literature on simultaneous contrast in the form of advice on how to produce strong contrast effects. For example, Rivers (1900) advocated contiguous surfaces without "differences of texture or of glaze," and Woodworth (1938) recommended "depriving the field of object character" in various ways. All these maneuvers tend to lead subjects to equate contrast. The HSD is just one convenient and particularly effective method.

5.2. Why Contrast Brightness? Analysis Into Objects, Lighting, and Spatial Configuration

What do all these methods have in common? It is doubtful that any generalization will cover all of them, because some, such as getting rid of outlines around the test patches, probably have their main effect on low-level processes of contrast coding, whereas others seem more likely to have their effect as the HSD does by preventing subjects from having a

---

[19]The reader should be clear that it is *comparative judgments* of two objects that are in question throughout this discussion (though one of the objects may be only remembered). Speaking of a single object as "having contrast brightness" generates problems, as we see later.

clear perception of the objective situation. Nevertheless, there is a hypothesis that will encompass most of them. It is suggested by the demonstrations by Gilchrist and Arend that local contrast can be overridden if one can see clearly that the backgrounds differ in reflectance rather than in illumination. That is one instance of the general process of parsing the scene into the three components of objects, lighting, and spatial layout.

The hypothesis is that judgments follow contrast brightness when the parsing into objects, lighting, and spatial layout does not come into play, either because it is not necessary because the objects are seen against the same background, or because it is in some way hindered. When it does come into play, judgments will follow contrast brightness only in special cases such as when only illumination differs between two scenes. This is only a slight development of ideas that can be found in Whittle and Challands (1969), Gilchrist et al. (1983), and other places in this book, and in earlier sources such as Woodworth (1938) and Evans (1943). Nevertheless, because it is not a familiar idea to many people, and because I have found it a surprisingly useful organizing principle that has made many observations fall into place, I think it merits a little more discussion.

Analysis of the scene into objects, lighting, and spatial layout is clearly not a single process except at a broad functional level of description. It involves, for instance, shape from shading, texture as a cue to illumination, the integration and classification of edges, and many spatial cues that are almost independent of color and lighting. But a broad functional level of description may be the right level for present purposes, because it allows us to understand what is happening, in a way that lower level descriptions such as integration or lateral inhibition do not permit. The parsing must be continually going on. Much of it is probably local (e.g., Barrow & Tenenbaum, 1986; Pentland, 1989), but there must be a global mechanism to achieve consistency over the whole scene. It must involve cortical mechanisms, and it is probably relatively slow (Sewall & Wooten, 1991). Although it is tempting to assume that this process is fundamental to all vision, in fact some complex visual behavior either does not need it or needs it only in a rudimentary form (e.g., reading).

The hypothesis accommodates all the phenomena we have been discussing:

1. The parsing should be irrelevant when there is only one background.

2. It is reasonable that in the kind of matching experiments we have been considering, the parsing should require that the difference in surround brightnesses can be clearly seen. When it cannot be, as in the HSD, the parsing is not activated, so judgments are determined by only local contrast.

3. In the experiments previously reviewed, it was the parsing of the

surround differences as reflectance or illumination differences that determined whether contrast or luminance was equated.

4. When different surrounds are visible, the parsing should be triggered automatically and so, for example, intrude into simultaneous brightness contrast experiments as unwanted fluctuations in attitude. It must be triggered automatically by low-level features because the perception of objects and space is its outcome not its precondition.

5. These "unwanted fluctuations in attitude" take the form of an awareness of alternative interpretations of the display. This is exactly how the parsing should be experienced.[20]

6. It is reasonable that for the parsing to work well, it should require not just mere visibility of the surrounds but what we call "a good look." It is a complex process that requires much information. Woodworth (1938) said: "Conditions which favour clear perception of the objective situation tend to annul contrast" (p. 570).

7. Most of the traditional recipes for maximizing contrast effects become comprehensible as ways of impairing the parsing. For example, having all the stimuli in one plane removes most of the spatial cues. Avoiding texture removes cues to surface character and direction of lighting. (It may also remove microstructure carrying local color information.)

8. The parsing is profoundly bound up with the perception of spatial layout, and we know this is important (e.g., Adelson & Pentland, 1990; Gilchrist, 1980; Nakayama, Shimojo, & Ramachandran, 1990).

9. If, as is likely, the parsing is relatively slow, it would be unable to keep up with flashing or flickering stimuli or surrounds, which would explain why they generate good contrast brightness (section 7.1 of the previous chapter). An alternative explanation would be that there is a dynamic dimension to the parsing—objects versus events—in which flashing or flickering lights are treated as changing illuminations and so yield contrast brightness for that reason.

10. Finally, this hypothesis suggests a partition of the processes underlying color perception, and therefore of the problem domain, that seems intuitively right: between adaptation and contrast coding at low levels, and the perception of objects and lighting at higher levels.

Can we now answer the question behind the chapter title: What is the

---

[20]Evans (1943) wrote: "A very interesting feature of all simultaneous contrast work and one which deserves more attention than it has received is the fact that the amount of induced colour is partly under voluntary control" (p. 594). Parrish & Smith (1967) obtained more simultaneous contrast in judgments based on "general impression" than with focal attention to the test patches. Was this how the Impressionists learned to see purple shadows in foliage?

relation of contrast brightness to ordinary seeing? My best guess is that it is always indirect. Contrast brightness reflects the working of the early stages. The contrast code is always potentially subject to transformation by later stages, although under many situations judgments of brightness and lightness follow contrast brightness relatively untransformed. Examples include a scene of constant reflectances with illumination changing over time; the perception of shadows and conditions generating high contrasts, such as seeing an object against the sky. We could also include judgments explicitly of apparent contrast because these are quite close to contrast brightness. Perhaps these situations are so common that they have access to a kind of default mode, requiring little higher level processing.

6. CONTRAST BRIGHTNESS: NATURE AND LIMITATIONS

6.1. The Phenomenology of Contrast Brightness

In the HSD we can see no illumination difference and so cannot choose between lightness and brightness. However, the other objective distinction subsumed by the lightness–brightness distinction (section 1.3) is relevant: Decrements look like surface colors, and increments look self-luminous. The object quality is determined simply by the luminance ratio and the edge quality. It may be the sharpness of the edges that make decrements look like pieces of paper rather than shadows. Over most of the range the observer has no choice between lightness and brightness, and even in the overlapping zone of low-contrast increments that are not clearly either surfaces or lights the ambiguity does not affect the matches. In this situation it is helpful, though imprecise, to use the words *brightness* and *lightness* for the two parts of the single bipolar dimension that I have called *contrast brightness*. It is Hering's black–white dimension extended to include self-luminous lights that becomes in Heggelund (1974, 1991) the "black-luminous" dimension. I think it is what Evans (1974) meant by "Brilliance," which subsumed "Grayness" and "Fluorence."[21]

There are two reasons why contrast brightness may be a useful concept: scientifically, for the study of visual mechanisms, and practically, for work with monochrome visual images. This discussion is not concerned with applications, so the practical value is not discussed at any length. How-

---

[21]The description *black–white* is not a good alternative to *contrast brightness* because it seems to omit self-luminous lights and because it sounds achromatic. The properties of contrast brightness are probably independent of spectral composition, as long as it is constant. In Whittle (1992) the equal-interval scales for yellow and for gray displays were of exactly the same form.

ever, contrast is the crucial intensity dimension in, for example, black-and-white photography or grayscale monitors, and the psychophysics of contrast brightness, particularly scaling and discriminability, is of obvious relevance. In visual science in particular, when using grayscale displays, questions often arise in passing to which the data in these two chapters are relevant. Almost all the material could have been presented as contributions to the applied psychology of working with such monitors. This extends to questions of concepts and terminology because in such work the need is often felt for a dimension that is less associated with real-world distinctions between lights and surfaces than are current definitions of brightness and lightness.

The main value of contrast brightness in this discussion is that it reflects the mechanics of vision.[22] It corresponds to the visually crucial dimension of physical contrast and makes possible the psychophysics of contrast brightness discussed in the previous chapter. Its bipolarity is perhaps reflected in the physiological polarity between on and off center cells. It is not a unitary dimension phenomenologically. The appearances it yields, with no cues other than contrast, are those of light sources, surface colors (or perhaps shadows if the edges are not sharp), and holes, not to mention the zone of invisibility around zero contrast and the possible division of the decrement zone into light and dark grays. In this heterogeneity it is like the red–green or yellow–blue dimensions, which also reflect visual mechanisms. These dimensions all have phenomenological significance, for instance, in implying the disjunction between red and green, but they would be supported only tenuously by the evidence of phenomenology alone.

It is therefore not surprising that the bipolar contrast brightness scale is a natural representation for some aspects of vision but not for others. *Positive brightnesses* are an appropriate description of increments that seem to give out light and are often made by adding light to the background. *Negative brightnesses* are appropriate to holes or to stimuli that are made by subtracting light. *Zero brightness* is a nice representation of invisibility or nonexistence. From other points of view the bipolar scale is unsatisfactory. Because brightness is apparent luminance its zero should correspond to black. And it is odd to regard gray surface colors as negative quantities with zero as their maximum. This recalls the difficulty mentioned in the previous chapter (section 2.4.1) of whether to describe decrements in terms

---

[22]The previous paragraph reminds us that it also reflects our laboratory apparatus, although not narrowly, as is shown by its 19th-century manifestations. Positive and negative contrasts are not found only on computer monitors. Similarly, earlier discussions of lightness in terms of amount of white reflected technologies of sector disks or of mixing pigments. We always hope that these are not so much constraints as happy marriages of technology and knowledge.

of darkness or lightness. The choice depends on several factors, such as whether we are seeing them as shadows or surface colors.

6.2. The Paradox of Taking Background Brightness as Zero

The most puzzling feature of contrast brightness is its relativity. It is doubly relative: with respect to its zero because it depends on luminance differences ΔL, and with respect to its scale because the difference signal is attenuated by a function of the absolute luminance. Here I consider the zero problem. It is puzzling because one would expect the zero point on our bipolar brightness scale to be in some sense at the surround brightness, because when contrast is zero that region of the visual field has the brightness of the surround. Yet in the matching experiments the background brightness is "discounted." The background has an adapting (attenuating) effect but does not otherwise affect either the brightness or the discriminability of superimposed stimuli. Whatever its own brightness, it can be taken as zero contrast brightness. This would not be surprising if subjects were consciously judging relative brightness (apparent contrast). What makes it puzzling is that in the matching experiments they took pains to base their judgments on the absolute brightness of the center of the patch.

It is as though once there is an *object*, and usually this just means once there is a closed contour, its edge-contrast determines the brightness of that region of the field. The contour creates a gestalt organization, a figure, and a ground. In that organization the surround brightness is treated as zero. The contradiction with the fact that brightness at zero contrast is the same as the background brightness suggests that there might be some kind of discontinuity as contrast passes through zero. What one notices if the transition is slow enough to see the object disappear and be replaced by ground is a switch to a new figure ground organization, which imposes a new reference brightness level. A great importance of outlines, which we often forget, is that they maintain such organization.

6.3. The Intransitivity of the Relation "Brighter Than"

Another perspective on the problem is provided by the observation that the relation "brighter than" is radically intransitive (Whittle & Challands, 1969). Figure 3.10 shows an instance of this. It represents the three lights A, B, and C marked in Heinemann's (1955) data in Fig 3.1. The luminances are drawn to scale, which makes it quite clear that A is brighter than B, and B is much brighter than C. The latter difference will be lessened by the dichoptic presentation, but still present. Heinemann's data, however, show

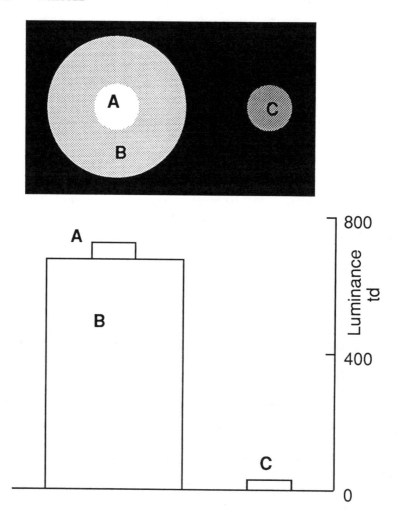

FIG. 3.10. The intransitivity of the relation "brighter than." The lower part of the diagram shows the luminances, drawn to linear scale, of the lights A, B, and C in Fig. 3.1. If $b(A)$ is the brightness of A, then $b(A) > b(B) > b(C)$ and yet $b(C) > b(A)$.

that C is much brighter than A. C is on a higher isobrightness curve, and the difference corresponds to more than a log unit of luminance in the dark. Such intransitivity is a simple consequence of brightness being a function of luminance ratios, and it is surprising that it is not more often remarked on, given how often the latter claim is made. It is obscured in the HSD because the haploscopic presentation means that B cannot then be seen to be brighter than C. But the same matches are made in side by side displays like Heinemann's where the relative surround brightnesses are clearly visible. There too the background is "discounted" and the consequent intransitivity shows up undeniably.[23]

This is one of those disturbing observations that the study of perception sometimes produces, like the fact that mirror images reverse left–right but not up–down. We find resolutions of these puzzles that seem perfectly satisfactory, but then a new point of view occurs to us and makes the problem as disturbing as before, like a conceptual impossible figure. So in this case I make a few remarks that I think are helpful, but I do not suppose that I have disposed of the problem.

A sentence from Merleau-Ponty's (1945) *Phenomenology of Perception* describes the situation well: "The visual field is that strange zone in which contradictory notions jostle each other because the objects . . . are not, in that field, assigned to the realm of being, in which a comparison would be possible, but each is taken in its private context as if it did not belong to the same universe as the other" (p. 6 of the English translation, discussing the Müller–Lyer illusion).

Such conflicts do not hinder ordinary perception. In "the realm of being"—the real world—we compare A and C along objective dimensions of reflectance and illumination, and no problem arises. But the difficulty does not come primarily from thinking about a "visual field" rather than objective properties, because it remains just as serious if we eschew all reference to brightness and talk only in terms of judging luminance. The problem seems more likely to be that we are not biologically adapted to judging luminance. We can judge the biologically relevant dimensions of surface color or illumination but not the scientifically defined quantity, luminance. Our visual system gives us access to relative luminance, but not absolute. To recover that, we have to either consciously compensate for the evolutionary bias of our senses, or, better, use instruments.

These logical puzzles are not peculiar to brightness. They arise for any relative variable whose relativity is repressed; that is, for which the reference level is taken for granted, kept out of awareness or current consideration. Perhaps most psychological variables are of this type. Consider warmth, or wealth, or happiness. In all such cases the problems arise when

---

[23]An algebraic formulation suggests the logic of the situation quite neatly. Denote the brightness of A by $b(A)$, and so on. Then we can say that $b(A) > b(B) > b(C)$, when the middle term $b(B)$ is "discounted," as background brightness is, implies nothing about the relationship of $b(A)$ to $b(C)$. The "discounting" of $b(B)$ seems to imply that it cannot validly be arranged in a series between $b(A)$ and $b(C)$ even though it can be unambiguously compared with each alone.

If you set this situation up, you may have to be content with the observation that A is only a little brighter than C, and somewhat ambiguously so, even though the other differences are large and clear. It is not so easy to get into the "attitude" of Heinemann's or Wallach's subjects. However, even the weaker demonstration implies intransitivity.

the reference levels are themselves brought into comparison. You might say that we simply make mistakes, but they are mistakes of great psychological interest. We have quantities related to local reference levels, which is what we usually need.[24] But when we wish to compare two quantities across different contexts, we must take their respective reference levels into consideration, which we often fail to do. In the case of luminance, the failure is built into our visual system, which "discounts" the background brightness. Nonpsychological examples include velocity or weight, for which the problems show up in space travel or relativity physics, when novel contexts require comparing reference levels.

Although the difficulty in our case does not come primarily from talking in terms of a visual field of brightnesses (sensations, qualia) analogous to the external field of luminances, that does compound the problem because it introduces an extra mystifying layer of concepts. For such an analogy to be useful, the relation "greater than" should surely be transitive for brightness as it is for luminance. Instead, "the visual field is that strange zone" in which brightness relations are intransitive. In thinking about brightness perception, we tend to assume that *brightness* is the main thing to be explained. We might be better advised to take the *act of judging* as the primary psychological reality. Perceiving is "judging" or "comparing" or "trying to decide" or "looking to see." These are just uncontroversial ordinary-language descriptions of what is going on. The conceptual status of "brightness" is more dubious. It is a quantity inferred by some vision scientists in a particular philosophical tradition because they thought judgments must be *of* something, and they would not rest content (although they should have done) with saying that the judgments are of the object properties reflectance, illumination, or luminance. An intervening layer of mental representations was postulated. The paradox of intransitivity is one among many arguments against its usefulness (see Dennett, 1991, for a more thorough philosophical justification for the primacy of judgments over qualia).

6.4. The Limitations of Contrast Brightness

I argued that the account of contrast brightness given in the previous chapter does not apply without qualification to ordinary vision. The intransitivity problem points to an equally severe limitation in its descrip-

---

[24]"Furthermore, what is so amazing about simultaneous contrast effects... is that the visual system seems to take them so seriously. That is, we get what looks like wrong answers in situations as simple as the Bernussi ring... where we would think that almost any sensible scheme would give an answer reflecting the objective truth of the situation. I find this so striking that I am tempted to believe that relative observations may be *all* one relies on" (Marr, 1982, p. 259).

tion even of "sensory" judgments of brightness. In the display in Fig. 3.10 the annulus B has two edges, an inner and an outer, so its contrast brightness is unknown. This is the problem of homogenization or filling-in discussed in the previous chapter (section 9.3.1). But the argument of the previous section implies that the problem is more than just finding the right value. There is literally nowhere on the contour map of Fig. 3.1 where a single-valued annulus brightness can logically go. I have marked it only as a luminance value on the abscissa. As a contrast brightness it would have to be both on a higher contour than C and a lower one than A, which is impossible. This shows again that the enterprise of representing all brightnesses in a visual field is incoherent. The map represents only a restricted class of judgments of contrast brightness.

The contrast brightness of B judged relative to C would be the point B_C on the ordinate (because its surround $L_b = 0$). Its contrast brightness judged relative to A is not defined. The contrast brightness story as yet says nothing about the comparison of adjacent luminances, the simplest kind of brightness judgment. It may depend on the contrast brightness of either the bright or the dark side or their apparent contrast, but we do not yet know. This is a serious limitation for an account of the dependence of brightness on edge contrast.

Consider for a moment where we might represent this contrast brightness of B judged relative to A. Because its (L, L_b) values are the same as those of A, but interchanged, a reasonable guess is that it would be the reflection of A in the diagonal, marked as the point B_A. B would then be represented by two quite distinct values, B_A and B_C, even though such an annulus usually appears to have one uniform brightness. This is a problem if we persist in thinking of all regions of the visual field as having brightnesses independent of our judging them, in the same way that they have luminances. If on the other hand we remember that a brightness is only the outcome of a particular *judgment*, the problem is reduced. On this view, what we call *sensations* are not the data ("sense-data") on which we base perceptual judgments but the answers that our perceptual systems give us when we use them to ask questions of the world. This view would, however, need to be supplemented by a theory of the preattentive visual field.

6.5. What Colors Are Seen in the HSD?

This is an obvious question to raise about the experiments on contrast color shown in Fig. 2.22 of the previous chapter. In the HSD subjects matched all homochromatic weak decrements, regardless of the monocular color (which was saturated red, green, or blue), with a gray. Now a weak decrement in a saturated red field does not normally look gray. It may look grayish by comparison with its surround, but not a completely achromatic gray. So has the color been changed by the HSD? (The same questions arise for brightness, but they are easier to talk about for color because we have names for colors.) There are various experiments that one can do to check on possibilities like binocular interaction, and they have not yet all been done.[25] But the interesting fact in the present context is that it is rather hard to answer the question in a simple phenomenological way. There is no problem in making the match. It is, as I said, subjectively very satisfactory indeed. While I am making it, I think that both patches look gray. But if I then close the eye that sees the gray display, the color of the other patch does not change suddenly. It is rather that after a while I may notice that it looks reddish again. If I then open the other eye, the patch still looks reddish, until maybe I try to critically check on the match, and then I see it as gray again. Now it may be that the processes involved are just rather slow (and again there are possible experiments), but I think that it is at least an interesting possibility that these situations may be better understood in terms of the idea that judging is primary and "having a color" is secondary. Such experiences are not specific to the HSD.

7. EXPLAINING SIMULTANEOUS BRIGHTNESS CONTRAST

This was the problem with which the whole investigation started.[26] It has turned out to be not one but many questions. A good place to start is with the opposition between on the one hand the thesis of the previous chapter, that simultaneous contrast is a huge effect, causing luminances that differ by a million to one to look the same brightness and representing one of the most fundamental processes in brightness perception, and on the other hand Gilchrist's (1988) evaluation of it as a minor failure of constancy, producing illusions that are only one or two Munsell steps in magnitude.

Clearly we are talking about different things. The previous chapter dealt with contrast brightness in the situations that produced it in the

---

[25]We obtained the same result for flashed stimuli in a side by side display where there was no possibility of binocular summation with the contralateral surround, and for the case of brightness Whittle and Challands (1969) concluded that such interaction was not important.

[26]It has disappeared from view, like the presenting problem in a psychoanalysis. As the inquiry proceeds other problems come to seem more interesting and fundamental. Nevertheless, readers, like friends and relatives, persist in wanting an answer to the original problem and deserve it because they have to put up with a lot along the way. But the answer when it comes is unsatisfactory, consisting partly of a lecture on why the question was not the right one, and partly of an inconclusive and overcomplicated attempt to answer it from a cognitive framework in which it has ceased to make much sense.

purest form. It showed the well-known Weber behavior of the visual system over a luminance range of 1:10^6 or more. *Simultaneous brightness contrast* in this context is just the name for suprathreshold measurements of this behavior for the special case of steady lights. This is not a new meaning of the term. Heinemann's (1955, 1961) data are often quoted as standard measurements, and we have seen that they agree well with mine. Gilchrist, on the other hand, is talking about the residual effect that can still be seen within the modest range of luminances afforded by reflecting objects under one illumination, and after the aforementioned integrating processes have done their work.[27]

Therefore the first step in explaining simultaneous brightness contrast is to distinguish different meanings of the term. If you want to know about contrast brightness, turn to the previous chapter. There contrast brightness was ascribed to a spatial filter plus gain controls and cortical filling-in. The visual system responds to luminance differences, attenuates them by a function of adaptation level, and extrapolates brightnesses across the regions between edges. Perhaps the most important component of this is the responding to luminance differences. The familiar concept of lateral inhibition, so heavily used in discussions of brightness contrast, was not explicitly mentioned. It may be involved in both the spatial filter and the gain controls, but explanations primarily in terms of lateral inhibition omit crucial levels of description. To explain brightness contrast in terms of lateral inhibition is like explaining the jerky progression of a learner driver in terms of the explosions in the cylinders of the car's engine. The explosions have a place in the causal chain, but regarding them as causes specifically of the jerks is to be mislead by a superficial analogy.[28]

If on the other hand you want to know about the modest illusion of lightness contrast that the integrative processes do not remove, then turn to Gilchrist's discussion. Explaining particular textbook demonstrations is probably not straightforward. They are usually pictures that do not unambiguously represent differences in either surface color or in illumination, and as has been argued, it is what they are seen as representing that is crucial, not what they are.

Everything I have said in these two chapters about simultaneous contrast has been restricted to center-surround displays with only a simple

---

[27]He was also talking of lightness rather than brightness contrast. The preceding discussion showed that what was crucial was that the *surrounds* were seen as differing only in reflectance, not the aspect of the test patches that was being judged. And in the HSD, decrements are effectively being compared in lightness, but the results follow contrast-matching.

[28]See Maffei and Fiorentini (1972) for a neurophysiologists' critique of explanations in terms of lateral inhibition. They went even further and wrote: "We have some doubt as to whether the organisation of individual receptive fields at any level in the visual pathway has anything to do with simultaneous contrast" (p. 71).

step in luminance or color between the two; that is, with no outline or wider gap. My assumption is that the center-surround configuration is a particularly important one, which will tell us about the contrast dependence of brightness and color, and I think that has been borne out. Yet the phenomenon occurs and has been frequently studied with spatially separate patches. I expect such situations to yield smaller effects, and the results obtained in them to be more difficult to analyze, but I do not pursue this topic here.

8. CONCLUSIONS

8.1. Summary

This chapter related the contrast dependence of brightness and lightness judgments demonstrated in the previous chapter (contrast brightness) to vision under more normal conditions. The argument started by considering the results of four studies that provided a halfway house between ordinary seeing and the restricted conditions under which contrast brightness is most reliably demonstrated. In all four studies brightness or lightness matches were made between patches in surrounds. The data agreed most closely with contrast brightness for *lightness* matches between patches in surrounds of the same reflectance but under different illuminations. When the surrounds had different reflectances, lightness matches were determined not by the luminance contrast of patch and surround but by the patch reflectance. When illumination was constant, this amounted to equating luminance in spite of very different edge-contrasts: We sometimes perceive surface colors with negligible distortion by simultaneous contrast. Some *brightness* matches followed contrast brightness but others deviated toward equating luminance. The whole pattern of results was interpreted in terms of different perceptual tasks, including (but not restricted to) two types of lightness constancy: with respect to changing illumination, with reflectances constant, and with respect to changing surrounds (background or foreground), with illumination constant. The intermediate results of some brightness matching experiments were attributed to the perceptual ambiguity of their stimuli.

If the signal from eye to brain is contrast coded, equating luminance implies something like integration by the brain. This was briefly discussed, and it was argued that some features of *assimilation* phenomena suggest that they may depend on processes involved in this integration.

The evidence reviewed showed that judgments of both brightness and lightness normally depend on parsing the scene into objects, lighting, and spatial layout. (Integration is one component of this.) I argued that judgments follow contrast brightness either (a) when this parsing is somehow hindered, or (b) in some special cases for which contrast brightness may be suitable without further transformation. The parsing requires conditions that favor clear perception of the objective situation, and there are therefore many ways of impairing it. In the experiments of the previous chapter it was prevented because the patches to be compared either were, or appeared to be, seen against a common background.

Finally, the discussion returned to the topics of the previous chapter: contrast brightness and simultaneous contrast. The heterogeneous phenomenology and intransitive nature of contrast brightness were pointed out. The intransitivity raises serious problems for the notion of a visual field of brightnesses. The problems are reduced if we take judgments as the primary psychological reality, not "sensations." Two senses of simultaneous contrast were distinguished, one a very large effect, one very small. It was argued that explanations in terms of lateral inhibition missed the point.

8.2. A Revised Partition, but not a Split, Into "Sensory" and "Perceptual" in Color Perception?

We have two centuries' accumulation of empirical and theoretical literature on the effects of adaptation and contrast on brightness and color. Bartleson (1977b) wrote of the literature on chromatic adaption alone, that there is now "an amount of experimental and theoretical information bearing on the subject so vast that it would fill an entire book even in cryptic summary form" (p. 50). A principal argument of these two chapters is that this large body of knowledge is clarified if one partitions the field in a particular way. On the one hand there is a relatively simple core structure underlying much of it: the behavior of contrast brightness and contrast color. These depend on the sensory processes of early vision. On the other hand there are the effects on brightness and color of the higher level perceptual processes that analyze the scene into objects, lighting, and spatial relations.

Such a partition accords with current intuitions. My arguments are too circular to prove its necessity. My claim is rather for its fruitfulness. It makes many old findings fall into place and suggests new lines of thought. The literature is full of telling observations and theoretical insights about this and that factor, from adaptation and local contrast to effects of spatial layout and the parsing into objects and lighting. Yet we repeatedly forget these lessons and keep hoping like alchemists for a single magic solution to a problem like color constancy. Then we gradually rediscover the complexity of the problem and that most of the complexities have been pointed out before. Such forgetting is often the sign of a split, a partition that has become a chasm. In this case there have been obvious splits both within and between the traditions of work that have thought of themselves as "sensory" and "perceptual." So although on the one hand we need the partition to bring order into the field, on the other we must not allow it to become a split, lest we have to recreate old knowledge yet again.

The arguments for a partition are strong. A whole tradition of work, including Hering, Mach, Wallach, Heinemann, Evans, Glad and Magnussen, and Heggelund, has drawn attention to the large and dramatic effects of local contrast on color appearance. These are such powerful phenomena and so strikingly independent of knowledge or expectation that when you see them well demonstrated it is hard to avoid the conviction that they must be of great functional importance. The conviction is reinforced by the argument that the dependence of color on stimulus ratios is just what is required to mediate the remarkable achievement of the visual system in making surface colors look much the same over an illumination range of 1,000,000:1. Many people have seen that these phenomena are aspects of light adaptation and the Weber behavior that it produces. (Though also, many have not. This has been one split: between studies of color appearance and sensory work on increment thresholds or contrast sensitivity.) My measurements illustrate this particularly clearly, and I argued that several different lines of evidence converged on the conclusion. Measured under appropriate conditions, brightness becomes just another contrast dependent response. There are many things we do not understand about contrast brightness and color, but in every case we can see promising lines of attack within the current framework of psychophysical and physiological knowledge of relatively early visual processes, from retina to striate cortex.

However, although this work shows the powerful effect of local contrast, another tradition shows that it can be virtually overridden. This is the work on lightness constancy by Katz, Gelb, Kardos, MacLeod, Hsia, and more recently Gilchrist and Arend. This work has been concerned with the appearance of real objects, and with the diverse ways in which we take illumination into account. Experiments often excluded the influence of local contrast at the outset, as an unwanted complication. (This shows another split, now between different traditions of work on appearance.)

Now, the evidence accumulated by both traditions is formidable. It is sufficiently strong to justify a working consensus like that allowing the coexistence of trichromatic and opponent stages in color vision. Both traditions can be accommodated within a framework such as has been proposed many times before, from Helmholtz to Marr, that admits both sensory and cognitive processes in the perception of color. The retina codes contrast. The brain generates our perception of surface colors, lighting, and transparent media on the basis of this contrast information, but always within complex context-sensitive perceptual processes.[29]

A partition into "sensory" and "perceptual" is hardly a novel idea. There are two things slightly unusual about this one in the contexts of brightness or color. The first is its insistence, in line with a more recent huge accumulation of literature (Graham, 1989), that the sensory processes deal with contrast. This does not need arguing in most sections of visual science. The second is its acceptance that even the elementary visual sensation of brightness depends on global context as well as local. Both of these imply, in different ways, that brightness is less well behaved than has usually been hoped. If we study its determination by the low-level mechanisms, that is, contrast brightness, then we can create a rigorous psychophysics. This is more coherent than we had thought, in that it promises better convergence of different methods, but it is also more puzzling, less like how we expected brightness to behave because of the paradoxes of relativity. If we study the higher perceptual levels, we find that brightness and lightness can be radically dependent on the global interpretation of the scene: another form of relativity. Both these aspects of brightness show the pitfalls of reifying it as though it was a physical quantity that had somehow slipped across the matter–mind barrier. It has more in common with meanings than we have wanted to believe, and if we want a physical analogy it should be with a relative quantity like velocity rather than an absolute one like energy.

---

[29]Historically, these ideas show a kind of reversal. Visual psychophysics started with the search for a simple relation between brightness and luminance. We have come to realize that even for brightness, contrast is usually a more important variable than luminance. The project of writing this book came from the belief that that still needed to be argued, because so much work on brightness was still, in Gilchrist's terms, caught in the photometer metaphor. In the previous chapter I made my contribution to that argument. Even there, however, there was evidence that luminance is important in the perception of high contrasts. In this chapter we have seen more such phenomena, some of them clearly demonstrated by Gilchrist himself. We now, however, see these as secondary to the contrast-related phenomena, and to be explained in terms of a transformation of the contrast code used by early stages of the visual system.

4

Surface Colors, Illumination, and Surface Geometry: Intrinsic-Image Models of Human Color Perception

L. Arend
Eye Research Institute
Harvard Medical School

1. INTRODUCTION

The early visual processes that encode local contrast in the image are critical components of lightness and color constancy (see Whittle, Gilchrist chapters, this volume). In the Arend–Blake model (Arend & Goldstein, 1987a; Blake, 1985a), a model of brightness perception in complex fields, brightness fields are computed by spatial integration of local contrasts. However, contrast encoding is not in itself sufficient to produce lightness and color constancy in natural scenes.

In this chapter I analyze the problem of visual computation of apparent surface color. The analysis has several parts. First I argue that our natural and man-made optical environments present challenges to color constancy that cannot be met by known sensory processes. The spatial gradients of light in the retinal image are produced by a tangled web of factors in the physical scene. The problem requires a complicated computation, of which early sensory processes are only the first stage.

Next I present a heuristic model of human surface-color perception. I present a brief sketch of a probabilistic theory of human perception. From a combination of ambiguous sensory data and assumptions about the structure of the physical world, the perceiver computes a mental model of the external world. I then describe in more detail the surface perception stages of the computation. My surface perception model borrows extensively from computational vision. It differs from previous human perception models by addressing questions of what quantities are explicitly represented in our perceptions and how they are computed. It also draws attention to theoretical problems that currently prevent clear interpretation of data on human surface-color perception.

I then return to the question of the contribution of sensory mechanisms to constancy of surface color. Departures of early visual mechanisms from ideal behavior constrain subsequent computations of surface colors.

Chromatic and achromatic issues cannot be entirely separated, but for simplicity I present my arguments mostly in terms of achromatic surface color. Some of the chromatic issues are considered at the end of the chapter (section 6).

2. THE PROBLEM

2.1. Single-Illuminant and Multiple-Illuminant Scenes

Arend and Goldstein (1987b, 1989, 1990) and Arend and Spehar (1993a, 1993b) found that subjects had excellent lightness constancy in multiple-illuminant scenes. The quantitative results agreed with the qualitative appearance of the patterns: Even in our very simple experimental world of matte, coplanar surfaces, the mapping from the retinal image to the perceptual representation was multivalued. In viewing our structured array of luminance patches, one perceives three distinct, superimposed spatial arrays; an apparent illumination distribution, an apparent reflectance (lightness) distribution, and a brightness distribution.[1] For reasons discussed later I hereafter refer to these multiple response arrays as *intrinsic images*.

This perceptual representation depends on the retinal image rather than on the physical objects producing the luminance array. We see the same intrinsic images whether looking directly at the monitor (38 light sources), a projected transparency of the monitor (38 illuminances, 1 reflectance), or a photographic print of the monitor (38 reflectances, 1 illuminance).[2]

Lightness constancy over multiple-illuminants in a single scene places

---

[1]It is possible that the brightness distribution is derived from more fundamental information, the distribution of brightness gradients. Recent measurements of brightness differences (Arend & Spehar, 1993a, 1993b) are described in section 5.2.2.

[2]This assumes that great care is taken to produce identical light arrays over the entire retina. This degree of retinal image control is not usually available. Usually differences in viewing conditions (e.g., light surrounding the display) produce detectably different appearances for these various media. The success of photography, however, shows that perceptual interpretation of images in terms of surfaces and illuminations is, to a first approximation, undisturbed by these differences. As the technical quality of virtual reality displays improves, we can expect that the sense of "presence" (i.e., the impression of actually being in the represented scene) will increase.

much greater demands on candidate constancy models than does constancy in single-illuminant scenes. In coplanar, matte scenes under a single-illuminant relative lightness constancy can be achieved by a system that preserves luminance ratios merely by normalizing the mean response by a factor that gives a constant value to a reference gray (e.g., white or middle gray). Such mechanisms are equivalent to von Kries adaptation. In achromatic images it is also a straightforward matter to obtain an illuminance image (in this case, spatially uniform) from this computed lightness image and the input image (if it is not lost in the lightness computation).

To produce surface-color constancy in multiple-illuminant scenes, on the other hand, a model must be able to analyze spatial luminance and chromaticity gradients into perceptual representations of their reflectance and illumination components (Gilchrist, Delman, & Jacobsen, 1983), a much more difficult computational task.

In natural 3D scenes homogeneous illumination is extremely rare due to shading, shadows, and secondary illumination of surfaces by light reflected from other surfaces. The illuminance on a curved surface decreases as it turns away from the normal to the light source. Objects cast shadows on other surfaces, with varying depths and degrees of sharpness. Objects in one illuminant partially occlude vision of objects in different illuminants. Neighboring surfaces become additional light sources as they reflect the primary illumination onto each other. The intensity of this secondary illumination depends on the surfaces' reflectances, and its spatial distribution depends on the surfaces' specularities. The complexity of these distributions of effective local illuminance is well known in the fields of image analysis (see e.g., Forsyth & Zisserman, 1990; Horn, 1977), illumination-engineering (Worthey, 1989; Wyszecki & Stiles, 1967), and computer graphics.

Even more complicated phenomena are also common—transparency, visible light sources, highlights, and reflections. A brief inspection of one's environment should convince the reader that all of these are frequently encountered and perceived.

2.2. Unasserted Color and Apparent Surface Color

Before discussing models, it is necessary to explain my usage of a few terms.

In the achromatic case there is fairly broad acceptance of standard terminology for the distinction between sensory color and apparent surface color, using *brightness* for the former and *lightness* for the latter. I use the following definitions, slight modifications of Wyszecki's (1986). Similar definitions have been offered elsewhere (CIE, 1970; Evans, 1974):

The *brightness* of an area of the visual stimulus is the apparent quantity of light emitted by that area. It may be considered to be "apparent luminance" or "apparent radiance."[3]

The *lightness* of a perceived surface area is the apparent quantity of light reflected by that area relative to that which would be reflected by a perfect reflector in the same physical position. It may therefore be considered to be "apparent reflectance."

For the chromatic perceptual dimensions there is no analogous widely accepted dual terminology. I define my own terms, for which I claim no particular virtues beyond simplicity, brevity, and no conflicting usage elsewhere. I collectively call brightness and its chromatic counterparts *unasserted color*, short for "perceptually unasserted color." Unasserted color is perception of the light coming from a direction in space (i.e., perception of the combined effect of illumination and reflectance). It is not attributed to surfaces or lights in the world and is in this sense preperceptual. I collectively call lightness and the chromatic dimensions of the apparent colors of surfaces *apparent surface color*, making no commitment at present as to how the chromatic dimensions are organized.

3. FAILURE OF UNIDIMENSIONAL CONSTANCY MODELS IN MULTIPLE-ILLUMINANT SCENES

Several models that have been considered to be surface perception models use algorithms that receive an image as input and return a single array representing surface color. These include classical "color appearance" models (e.g., Judd, 1940) and gradient-integration models, (e.g., Land, 1986). None of these models accounts for data on surface appearance in multiple-illuminant scenes.

Traditional sensory-process models of color appearance are incomplete as models of surface-color constancy. The sensory processes of these models produce a single array of apparent colors (see Shapley & Enroth-Cugell, 1984, for a review of achromatic adaptation, and Wyszecki, 1986, for a comprehensive review of color appearance). With careful design, algorithms modeled on simultaneous contrast and slow adaptation of achromatic sensitivity can generate responses that are approximately invariant under changes of scene illuminance, but in so doing they are usually

---

[3]The user may freely choose between apparent luminance and apparent radiance on the basis of convenience in the immediate context. Which is chosen determines whether the modifier "apparent" is considered to include or exclude the visual spectral sensitivity that constitutes the distinction between luminance and radiance.

4. SURFACE COLORS, ILLUMINATION, AND SURFACE GEOMETRY   163

considered to discard the illuminant information (e.g., Cornsweet, 1970, p. 379). The visual system clearly does not discard all illuminant information. In Arend & Goldstein's (1990) multiple-illuminant Mondrians, the lightnesses were illumination invariant, but the brightnesses varied substantially with illumination. The gradients of illumination over the patterns were perceptually represented as gradients of apparent illumination.[4] Sensory models are designed to predict the unasserted color of the light that comes to the eye (as a result of interaction of reflectance, illumination, and orientation), in that they say nothing about perceptual representation of surface and illumination properties. They provide no mechanisms capable of distinguishing illumination gradients from reflectance gradients.

Gradient-integration models are similarly limited. The limitations apply to all the edge-integration models as well, including the Arend–Blake, Retinex, and Horn models. For simplicity I illustrate here using the one-dimensional model from Arend & Goldstein (Fig. 4.1). In this model the relationship between the brightnesses at two points A and B is given by integration along a path connecting the points. First the logarithm of luminance is calculated for all points. Then the component of the gradient of log luminance in the direction of the path is computed and a threshold process is applied. The brightness at B, relative to that at A, is given by the line integral of the thresholded gradient components along the path from A to B.

Granted several assumptions about sampling of paths among points, the 1D algorithm does produce illumination-invariant relative lightnesses within uniformly illuminated regions of Mondrian scenes (i.e., the models' predicted values for patches are equal if the patches have equal reflectances; Fig. 4.2a). On the other hand, in scenes with multiple illuminations, like those simulated in Arend & Goldstein's (1990) unevenly illuminated Mondrian experimental displays, patches lying in differently illuminated regions have equal computed values when their luminances are equal (Fig. 4.2b), not when their reflectances are equal.

This dramatic failure in multiple illuminant scenes arises because the gradient-integration algorithms have only one potential means of analyzing luminance gradients into reflectance and illumination components: the threshold for spatial luminance gradients. All suprathreshold luminance gradients are added into the integral. This is not sufficient to account for the Arend and Goldstein data, which show that observers have excellent lightness constancy in the presence of visible illumination gradients.

---

[4]The gradient of illumination over Land and McCann's achromatic Mondrian was also easily perceived but did not prevent lightness constancy (J. McCann, personal communication, May 16, 1989).

LOG: $L(x,y) = \log H(x,y)$

DERIV: $E(s) = \vec{\nabla} L \cdot \vec{t}$

THRESH: $D = \psi E$

LINE INTEGRAL: $\int_A^B D = $ Brightness at B, relative to brightness at A

Problem: Value of integral is generally path-dependent due to threshold.

FIG. 4.1. Schematic of the one-dimensional version of Arend and Goldstein's (1987a) brightness model. Arend and Goldstein rejected this model (and all related one-dimensional models) in favor of the more appropriate two-dimensional version of their model (it is used here for illustrative purposes only).

These gradient-integration models have often been taken to explain perceived surface color, but they actually predict unasserted color. They assume perceived color to be single valued. The single value they predict corresponds well with surface color in single-illuminant scenes (coplanar surfaces with uniform illumination). However, once the scene is complicated enough that some region has obviously different lightness and

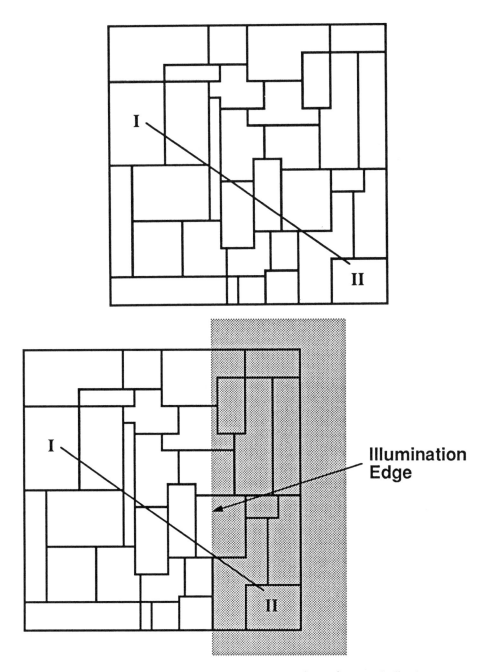

FIG. 4.2. (a) Retinex path for comparison of two patches within a single illuminant. (b) Retinex path for comparison of two patches separated by a steep illumination gradient.

brightness, it becomes clear that the gradient-integration models, like the color appearance models, predict unasserted color, not apparent surface and illuminant colors.

3.1. The Scene Segmentation Problem

All these unidimensional models share a flaw: a single global compensation for illuminant color cannot be correct for two simultaneously visible regions with different illuminations. To avoid this problem one might propose that such scenes are first perceptually segmented into several regions, each with its own homogeneous illumination. The algorithm could then be applied separately to each.

There are two problems with this proposal. First, Arend and Goldstein (1989, 1990) studied lightness and brightness in patterns that had no homogeneously illuminated segments (i.e., achromatic Mondrians that were completely covered by shallow illumination gradients). The data were very similar to those from the two-Mondrian experiment (Arend & Goldstein, 1987b). Subjects' lightness matches showed excellent lightness constancy even though the illumination gradients were clearly visible.

Second, to get the homogeneous regions the model must provide algorithms that can do the segmentation. Separating the illumination gradient between the segments from reflectance gradients is not trivial computationally. It constitutes most of the constancy problem.

4. INTRINSIC-IMAGE MODELS

The preceding arguments indicate that we need to understand the concepts and data of unasserted color as part of a larger scale model, one that provides for simultaneous perceptual representation of unasserted color, illumination, surface color, and surface geometry. The sensory-process concepts remain important in that they relate perception of unasserted colors in complex scenes to psychophysical and physiological research based on simpler stimuli and well-established sensory research paradigms, but they are only the first step of surface perception.

The larger scale model is needed for two reasons. Lighting variables are not just unstructured impediments to accurate perception of surface colors. The complicated interconnections among the multiple dimensions of surface perception are crucial to reliable visual computation of not only surface color but also surface depth and orientation. Apparent surface color can be adequately understood only through simultaneous consideration of these other visual computations.

The larger scale model is also needed as a heuristic aid in designing experiments. The Arend and Goldstein surface-color experiments have shown that observers can reliably distinguish between the perceptual dimensions brightness and lightness, and that the two have very different quantitative structures. The corresponding distinction in chromatic color has also proven useful (Arend & Reeves, 1986; Arend, Reeves, Schirillo, & Goldstein, 1991). There are several remaining problems in interpreting these data that will only be solved in the context of a larger model.

My proposed model of surface perception is only one part of an overall model of perception that seems to be taking shape in my work and that of several other contemporary perception researchers. It might help orient the reader if I briefly sketch the overall model before launching into the details of surface perception. It is not possible in this chapter to discuss all the details of the overall model and the arguments behind it, but many of them emerge in the details of the surface perception model.

Perception is computation of a mental model of the external world to serve as a guide for behavioral decisions. The objects in the mental model are human-scale inventions of the mind, constrained by the external world, but not themselves part of external reality. The rays are not colored; the colors are in the mind. The sensory data driving the mental model are inherently ambiguous with respect to their physical causes in the viewed scene. Nevertheless, the visual system automatically derives an interpretation that is highly probable (under most viewing conditions). It accomplishes this by supplementing the sensory data with assumptions and knowledge about the physics of the external world. This unconscious, automatic interpretation process is closely akin to Helmholtz's "unconscious inference." Computation of the perceived properties of surfaces in the scene is an early stage in this construction of the mental model.

A number of researchers in both human and machine vision have previously recognized the need for a theory with multiple intrinsic images for perceptual representation of surfaces and lights. I borrow the framework for my human model from computational vision rather than previous human vision work to take advantage of the more explicit structure of the computer models.

4.1. Computational Models

Our main concern is human perception. With respect to machine vision this means that the impressive robotic vision systems designed for restricted environments (objects and lighting conditions) are largely irrelevant. We need algorithms that perform successfully in a mobile visual system operating in general human environments. This has proven a very difficult computational problem. Ten years of attempts to develop algorithms for machine vision with the necessary capabilities have produced progress but are still unsuccessful. Human vision is so far the only existence proof of the possibility of successful surface perception in a general-purpose visual system.

Zucker (1987) characterized several artificial intelligence models of surface perception as the "second paradigm" of computational vision. Their distinguishing feature is a stage in which general-purpose information about surfaces (reflectance, orientation, depth) in the scene is derived from input image structure, prior to and independent of symbolic representation of objects and events. This intermediate computational problem, sometimes referred to as *middle vision*, includes the surface-color constancy problem.

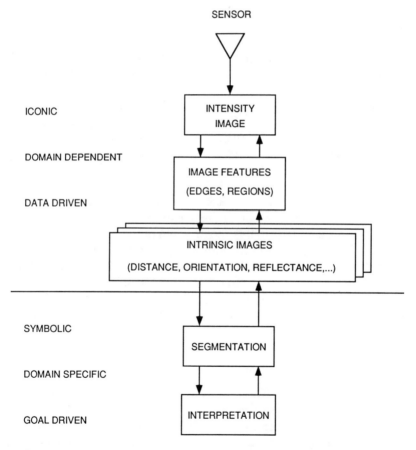

FIG. 4.3. Barrow and Tenenbaum (1978) model for computer visual system. Reproduced by permission.

Marr (1978, 1982) and Barrow and Tenenbaum (1978) made major contributions to middle vision theory. The arguments in this chapter have been strongly influenced by Marr's insights, especially regarding the emphasis on representation issues and computational demands. As a starting point for a human model, however, I have found the detailed structure of Barrow and Tenenbaum's model more convenient.

Barrow and Tenenbaum's (1978) full model is diagrammed in Fig. 4.3. The first stage is a set of computations that correct for sensor properties and identify local structures (e.g., edges and lines) in the input image, prior to any interpretation in terms of scene objects or surfaces. The result is an augmented intensity image.

The next stage, which produces the first scene-domain (as opposed to image-domain) information, is the one of greatest interest here. A set of arrays, called *intrinsic images*, are computed in parallel. The values in these arrays represent the machine's computed knowledge of the surface and illumination properties in the external scene that are physically responsible for the observed image intensities. The higher level symbolic processes in the model are of no interest for our immediate purposes.

4.2. Inverse Optics

Computation of the intrinsic images is illustrated in Fig. 4.4. The computation involves operations on image information that are analogous to inversion of the photometric equation. The physical process determining the intensity at a point x in an achromatic input image is described by

$$L = E R \phi \tag{1}$$

where L is image intensity, E is the local incident illumination intensity, and R is the local surface reflectance (all at x). The factor ϕ is a complicated function of the photometric geometry of the scene.[5]

To recover the reflectances of the surfaces from image intensities it is necessary to also simultaneously recover photometric geometry (e.g., sur-

---

[5]This is sufficiently detailed to illustrate the process for the immediate discussion. In the preceding discussions I tacitly included the geometrical factors in the illuminant distribution and talked about the equation as though it were $L = E R$.

For natural scenes with chromatic surfaces and lights, the equations describing image formation are much more complicated (Forsyth & Zisserman, 1990; Horn, 1977). Additional factors include multiple light sources of different sizes, types, and locations; surface roughness and specularity; and dependence of most of the terms on wavelength. The problem is restated later (equation 4) with some of the dependence on wavelength included. To know surface colors the human visual system must be able to take these additional factors into account, at least approximately.

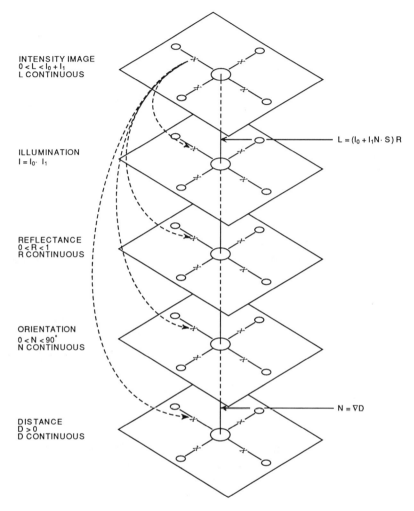

FIG. 4.4. Intrinsic-images computation (Barrow & Tenenbaum, 1978). Reproduced by permission.

face orientation) and illumination. This "inverse optics" computational problem is underdetermined by the information in the input intensities. There is too little information in the intensity signal to uniquely solve for the unknown optical quantities in the scene. To obtain a solution, the visual system must supplement the input image intensities with assumptions about the world it is viewing. The inverse optics problem and candidate solutions have been central issues in the field of image analysis (Barrow & Tenenbaum, 1986; Horn, 1977; Marr, 1982).

The Arend and Goldstein two-Mondrian stimulus provides a good example of the inverse optics problem. With sufficient attention to details, *exactly* the same pattern of retinal illuminances could be produced by an image of the monitor (65 light sources), a projected transparency of the monitor (65 illuminances, 1 reflectance), or a photographic print of the monitor (65 reflectances, 1 illuminance). To the observer, the display appears to be yet another physical arrangement, two identical arrays of 32 reflectances under two different illuminants. Because these diverse physical objects produce exactly the same information on the sensor array, there is logically no way for the visual system (machine or human) to veridically deduce the physical scene from this single view.

Given this ambiguity the best the perceiver can do is to base behavioral choices on a calculated best guess about the physical cause of the image. This suggests the following definition of a *percept: The subjectively most probable subjective physical explanation for the current pattern of sensory stimulation.*

Several components of this definition need elaboration. *Subjective* physics refers to the rules and entities that the visual system uses in its automatic computation of the mental model. The high success rate of human perception indicates that the rules and entities of subjective physics are sophisticated and orderly, and that they are closely related to the laws of "rational" physics, the instrumental science taught in school. Nevertheless, the two are distinct.[6] Subjective or psychological physics is largely beyond the reach of conscious, rational thought. Teaching me that Mach Bands are an illusion does not prevent my seeing them. Although interesting arguments have been made that experienced scientists "perceive" such rational-physics entities as electrons, bacteria, gamma radiation, and DNA, I am not convinced that they can enter "the picture in the head" with the same immediacy as, for example, the surfaces of visible, human-scale objects.

The *explanation* is not merely an abstract list of objects in the scene and their interrelationships, as is sometimes supposed in top-down perceptual models. It is an elaborated state of the mental model, complete in all the perceptible details of the scene. It includes specific values for distances, orientations, and colors of surfaces at all visible spatial scales, positions and colors of light sources, and so on. This need not involve vast numbers of degrees of freedom. Most of the information is not contingent. Rather, it

---

[6]Another reason to suppose that the two kinds of physics should be closely related is that rational physics is a structure built by human minds on a basis of orderly perceptions. In fact, the theoretical physicist, David Bohm (1965), argued that quantum mechanics is a kind of perception.

is entailed by the recognized physical structure, the relationships among the variables of subjective physics.

The second use of *subjective* refers to the probability computation. The probability that the sensory array was caused by any particular configuration of physical states depends on both the likelihoods of occurrence of the various states and the conditional probabilities that the observed sensory array occurs given the various states. In both cases it is the visual system's subjective probabilities that are involved, not the actual physical probabilities. Furthermore, as with subjective physics, there is no guarantee that the likelihood computation follows the laws of rational probability (e.g., Bayes rule). Once again, however, the rarity of serious illusions (erroneous elements in the mental model) indicates that the computation corresponds well to actual environmental contingencies.

Finally, *current* is not intended to deny that the computation and interpretation can depend on the context of preceding patterns of stimulation.

In Barrow and Tenenbaum's algorithm the visual system had two types of assumption regarding physical structure. *Within* each intrinsic image there were continuity and range constraints, and *between* intrinsic images there were constraints corresponding to photometric laws. To obtain unique solutions Barrow and Tenenbaum's constraints were very tight and closely tied to their artificial laboratory world. Neither humans nor machines can successfully use Barrow and Tenenbaum's particular assumptions in viewing general scenes. However, the distinction between the two kinds of constraints is useful. More realistic constraints are discussed later.

4.3. Human Perceptual Models

In work on human perception of surface colors, the first discussion of multiple representations appears to have been Katz's (1935). Gestalt psychologists devoted a great deal of attention to the problem (see Koffka, 1931, 1935, for reviews). Koffka was well aware of the distinction between lightness—which he called "whiteness" (1935, p. 243)—and brightness, and of our simultaneous perception of both.

Beck (1972) recognized that sensory models are incomplete, but his suggested second perceptual stage was described in rough terms only. In his model varying perceptual organizations produce different interpretations of sensory data, but there was no account of how these "schemata" and "traces" are established, selected, and applied. Whereas some of his organizational principles were described in terms consistent with eventual mechanistic description, he seemed to give at least equal weight to processes dependent on individual experience. He specifically rejected the concept of a quantitative linking of perceived illumination, perceived orientation, and perceived reflectance analogous to the physical link in the photometric equation.

More recently, Gilchrist (1979; Gilchrist et al., 1983) proposed that spatial arrays of edge ratios are sorted into reflectance and illumination edges prior to being integrated into parallel reflectance and illumination percepts. Bergström (1977; Bergström, Gustafsson, & Putaansuu, 1984) proposed that reflectance, illumination, and three-dimensional form vectors are analyzed from the luminance image by processes analogous to Johansson's (1976) vector analysis model of motion perception. Bergström and his students also provided a number of vivid demonstrations of the importance of perceived orientation in surface-color perception. Several other perception researchers have made less detailed arguments in favor of perceptual representations that resemble intrinsic images (Evans, 1974; Kozaki, 1973; Lie, 1977; Whittle & Challands, 1969).

All these efforts are important and address the problem in appropriate terms, but none fully answers our theoretical and empirical needs. The most developed efforts, Gilchrist's and Bergström's, lack an explicit, detailed representation of the inverse optics computational dilemma: how to analyze a single array of luminance gradients into reflectance, illumination, and surface-and-lighting orientation components. It is clear from our perceptual experiments, Bergström's, and others', that humans manage at least an approximate solution of this problem for most scenes. It is equally clear from research in image analysis (e.g., Horn, 1977) that the required algorithms are inherently complicated due to the wide variety of physical factors in image formation.

We do not yet know enough to specify the algorithms of the human visual computation. In the meantime, we need a heuristic model, one that outlines the perceptually represented quantities and the computational problem in sufficient detail to guide research.

5. TWO-STAGE HUMAN MODEL

For this purpose it is convenient to modify part of Barrow and Tenenbaum's model from computational vision (Fig. 4.5). I offer this preliminary model for heuristic and organizational purposes only; I make no claims that it has any explanatory power at this stage of development. It does, however, place greater emphasis than previous perceptual models on the computational problem of the second stage. Most of the concepts involved have been discussed in some fashion in the computational vision literature. To my knowledge, however, it is the first serious attempt to present a detailed intrinsic-images model of human surface perception.

The model consists of a sequence of stages (Fig. 4.5a), the first two of

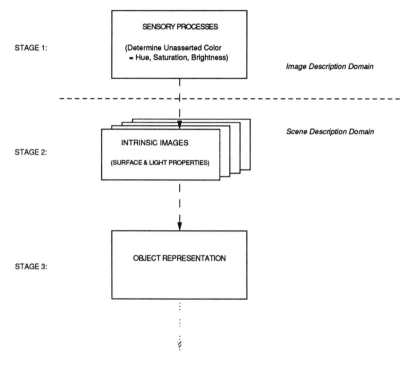

FIG. 4.5a. Continued

which are of interest here. Stage 1 consists of all sensory processes that affect the final color array of the image-description domain. All descriptions at this level are in terms of unasserted colors in image-centered coordinates (i.e., there is not yet any perceptual assertion about the physical causes of the image). Stage 2 (Fig. 4.5b) consists of analysis of the final image-domain color array into the first perceptual representation of surfaces and illuminations, properties of the scene domain.

5.1. Stage 2, Inverse Optics Processes

The inverse optics problem is insoluble without additional constraints supplied by the visual system, whether human or machine. These constraints are assumptions that the visual system makes about the photometric physics of the world it is viewing. Computationally oriented experimental work on human vision is in its infancy. For the most part we can only point to various computer-vision algorithms as examples of the kind of computation that must occur and list some of the more promising constraints to be investigated.

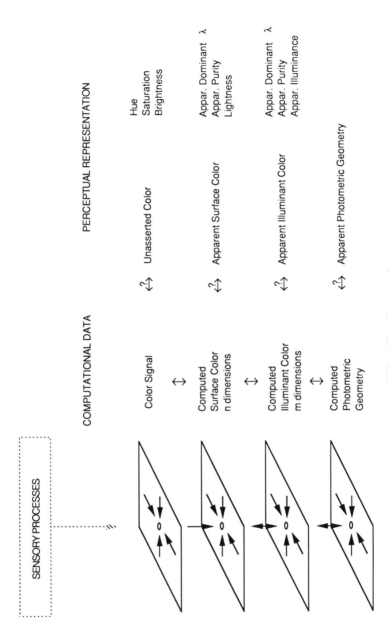

FIG. 4.5b. Continued

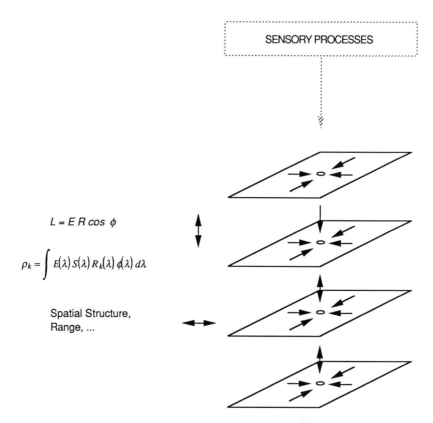

FIG. 4.5. Human intrinsic-images model. (a) Full model. (b) Surface perception (intrinsic images) computations and associated visual variables.

5.1.1 Between-Image Constraints

The Image-Intensity Equation. The relationship among the physical quantities represented by the intrinsic images is specified by Equation 1. In terms of spatial gradients the relationship is

$$\nabla \log L = \nabla \log E + \nabla \log R + \nabla \log \phi. \tag{2}$$

For surface perception to adequately represent the actual physics of surfaces, the relationships among the computed intrinsic images must approximate the simple equation

$$L^* = E^* \cdot R^* \cdot \phi^*, \tag{3}$$

where the starred quantities are the computed intrinsic image counterparts of the quantities in the external scene. In machine visual systems these computed values are explicitly known. In human vision they are not directly measurable.[7] We discuss this problem in more detail later.

Interdependence of Intrinsic Images. Marr's (1982) version of Mach's card illusion (Fig. 4.6) provided evidence of the linkage between surface variables that is expressed by Equation 3. When the dim area is seen as the shadow of a vertical white face, the apparent underlying reflectance is white. In this case the dimness is attributed to lower illuminance, consistent with the apparent 3D optical geometry. When the figure reverses in depth, the apparent geometry is changed. Now the brightness difference is attributed to a difference of surface reflectance. One sees a wedge-shaped object with gray surfaces. For some viewers this is a somewhat subtle change. The critical aspect to attend to is whether or not the dimness "belongs to" or "is a property of" the surface.

The chromatic version of Fig. 4.6 has analogous appearance. In the stimulus the dim area is made bluer than the bright area, as if there were two light sources, a fluorescent source located to the left and an incandescent illuminant to the right. When the blue area appears to be a vertical face, it appears to be casting a shadow with respect to the incandescent source. In this case the bluer unasserted color is attributed to shading. In the reversed appearance the apparent geometry is consistent with uniform illumination. The bluer area appears to be a wedge with a desaturated blue surface color. In the former case the blueness appears to be transparent. In the latter case it appears to be opaque.

This interdependence of the surface dimensions places a strong limitation on the modularity of their computation. I argued earlier that gradient-integration models cannot compute illumination-invariant reflectances because they include no mechanisms that can parse luminance gradients into reflectance and illumination components. This criticism can be generalized to include all previously proposed modular computations of intrinsics that require fixed or known values for the nontarget variables.

For example, "shape from shading" algorithms have the serious limitation that gradients of surface reflectance must not contribute to the luminance gradients in the input image. This requires that either the surface reflectance is perfectly uniform or, equivalently, that the reflectance gradients are known in advance so they can be divided out of the input image. There are a few special cases of scenes where surface color is in fact nearly

---

[7]There is no evidence at this point as to whether the intrinsic images contain gradients or scalar magnitudes after spatial integration of gradients. The measurable psychophysical variables are scalar magnitudes. Phenomenal surface colors are absolute, filled-in arrays, not spatial derivatives.

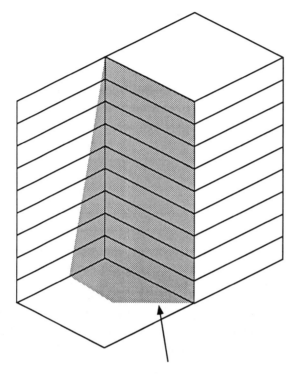

$$\nabla B \Leftrightarrow \nabla Appar.\ E,\ \nabla Appar.\ \phi$$
$$\nabla B \Leftrightarrow \nabla Appar.\ R$$

FIG. 4.6. Marr's version of the Mach card illusion. The perceptual interpretation of the brightness gradient changes when the figure's depth reverses. In one depth relation it is a gradient of apparent illumination associated with a gradient of apparent surface orientation. In the other it is a gradient of apparent reflectance (lightness).

uniform (e.g., surfaces of some celestial objects). In most cases, however, reflectances are not uniform, and the algorithms can work only if the image is first passed through a competent surface-color-constancy algorithm that is surface-orientation invariant. There is currently no such algorithm.

The same criticism applies to surface-color algorithms that require a Mondrian world. By restricting the domain to coplanar Lambertian surfaces, one asserts that the problem of discriminating luminance gradients caused by orientation gradients from those caused by reflectance gradients is someone else's. As in the orientation case, no such algorithm exists.

It appears that simultaneous solution for all the intrinsic images is unavoidable for general scenes. This imposes an additional criterion on candidate algorithms. Given otherwise equally successful algorithms for one of the intrinsics, higher preference should be given to the one most compatible in cooperative computation of all the intrinsics.

Beck (1972) specifically rejected perceptual models obeying Equation 3, citing several experiments he considered inconsistent with it. He argued instead that perceived illumination and perceived reflectance (and their environmental validities) could vary independently. There is, however, no hope of lightness constancy from algorithms inconsistent with the multiplicative relationship of Equation 3. It is not required that the subjective magnitudes of perceived luminance, perceived illuminance, and perceived reflectance scale linearly with the physical quantities, but the structure of the physical relationship must be preserved. Subjective luminance (brightness) does vary as a function of both illuminance and reflectance (Arend & Goldstein, 1987a). Everyday experience and Arend and Goldstein's data show that humans have excellent knowledge of relative reflectances in sufficiently complex scenes. Experimental evidence inconsistent with Equation 3 can possibly be attributed to experimental displays too impoverished to support human constancy algorithms, but there is currently no systematic evidence for this view.

5.1.2. Within-Intrinsic-Image Constraints

Distributions of Reflectance and Illuminance. A variety of sensory mechanisms have been proposed to automatically analyze luminance gradients in the input image into separate illuminance and reflectance images. At the beginning of this chapter I criticized their utility as color constancy mechanisms in terms of their single-valued response. There is, however, a second important criticism. They must frequently fail for general input images because they are based on a faulty assumption about the photometric physics of natural scenes.

Previously proposed color constancy mechanisms are based on the assumption that illuminance varies more gradually over space than reflectance.[8] This is a "within-image" constraint, an assumption about the statistics of the spatial structure within the illuminance and reflectance intrinsic images. In some cases this assumption has been explicit (e.g., Horn, 1974; Land & McCann, 1971; Maloney, 1985). In others it is implicit (simultaneous contrast, global adaptation). For simultaneous contrast to offset the effects of an illumination change on the color of a target region, the illuminants must also fall on surrounding regions. For global adapta-

---

[8]I am primarily referring to visual models, but image processing algorithms (Oppenheim, Schafer, & Stockham, 1968; Stockham, 1972) have been proposed that are also based on this assumption. My criticism applies equally to the image-processing algorithms. It suggests that their utility is limited for general input images.

tion to compensate for a change of illumination, large regions of the scene must be homogeneously illuminated. This is required to ensure that the local retinal light exposures, integrated over a time long enough to include several saccadic eye movements, are equivalent to steady viewing of a gray patch under the test illuminations.

A few mechanisms require that local spatial gradients of illumination must be shallower than local reflectance gradients (e.g., the threshold in Land and McCann's, Blake's, Horn's algorithms) if they are to be interpreted as color constancy algorithms. Most, however, require only that the support for the illumination distribution in the spatial frequency domain is at lower frequencies than that of the reflectance distribution.

This assumption about the ecological optics of natural scenes is probably unwarranted. Arguments for its validity have consisted of vague allusions to optics and appeals to intuition. In my opinion, arguments against the assumption are more compelling. The global spatial distributions of reflectance and illumination should be very similar due to shading and indirect illumination. Both natural and man-made environments consist of nested domains, with the texture at one spatial scale being form at a smaller scale. From across a large field, a forest is a form, with tree trunks and shadows of branches as texture. At closer range the forest is a region filling much of the visual field, with individual trees as forms and bark patterns as texture. Still closer the tree trunk is a region with bark patterns as forms.

At each scale most of the energy in the spatial distribution of reflectances is due to extremal and occluding edges, the boundaries of objects. At each scale, the shadows cast by objects are roughly the same size as the objects themselves. On a clear day, the edges of shadows have edges as sharp as the casting objects. The source aperture must be very large to make shadow sizes significantly different from object sizes. Objects under one illumination at one distance from the observer frequently overlap more distant objects that are differently illuminated. The result is an edge where illumination and reflectance covary. A less obvious reason for the spatial distributions of illumination and reflectance to be similar is secondary (reflected) illumination. Secondary illumination is strongest near the reflecting object. When viewed from a distance, large gradients of secondary illumination are therefore also likely to be on roughly the same spatial scale as the reflecting objects.

If this nested domain argument holds, then there is power in both the reflectance and illumination distributions at all spatial frequencies. Constancy mechanisms that rely on separate spatial support for reflectance and illuminance will fail.

It might be possible to rescue these algorithms by taking the nested domains into account, but it is not easy. It is not sufficient merely to bandpass filter the image into a pyramid of spatial scales. The reflectance and illuminance distributions are still confounded within each subimage. The algorithm must somehow evaluate spatial structure at each spatial scale as potentially being reflectance modulation within illumination modulation at larger scales.

Range Constraints. A second kind of within-intrinsic-image constraint concerns the permissible range of values. If the range of luminances and chromaticities in an image exceeds the gamut that can be produced by real reflectances under a single illuminant, one can reliably deduce the presence of multiple illuminants.

It is easily demonstrated that the visual system uses range constraints in complicated ways, if at all. We often perceive multiple illuminants when all the luminances in the image lie within the range of reflectances under a single illuminant.[9] Photographs of natural scenes can give a compelling impression of shadows and other illumination gradients. Rembrandt and other Dutch painters were famous for their ability to represent illumination differences for dramatic effect. Gilchrist and Jacobsen (1983) showed that lightness constancy is good even when a scene is viewed through a veiling luminance that drastically reduces the scene contrast.

5.2. Stage 1: Role of Sensory Mechanisms in Surface Perception

Let us now reconsider the contribution of sensory processes to constancy in the context of this two-stage model. The model suggests, as a hypothesis, that computation of apparent properties of surfaces (i.e., the intrinsic images) is based on image data that have already been preprocessed by early sensory mechanisms. As with Barrow and Tenenbaum's first stage, the hypothetical role of these mechanisms is to accurately encode local image contrasts by compensating for sensor limitations and perhaps also to encode primitive pattern characteristics such as edges. If so, any distortions of the input information that arise from less than ideal performance of the sensory mechanisms should be detectable as distortions of surface

---

[9]There is considerable variation in various investigators' measurements of the range of physically realizable reflectances. For example, Eastman Kodak (1986) gave ranges of 75:1 to 100:1 for glossy photographic prints and 40:1 for black ink on white paper. Halftones on glossy stock are given as 25:1 and on newsprint can fall to 5:1. Our original 19:1 range came from measurements of black-and-white art papers in our laboratory. We have recently been using 30:1 as a compromise value.

perception. In this section I discuss some distortions that seem likely given our current knowledge of sensory processes.

Because we scan our environment by moving our eyes, it is necessary to classify processes by how rapidly they work (Judd, 1940). *Fast* processes are those with time constants on the order of 0.5 second or less. Fast processes operate on the same time scale as the eye movements involved in scanning the scene. They therefore interact strongly with eye-movement-generated temporal modulation on the retina. *Slow* processes are those with time constants on the order of tens of seconds to minutes, slow enough to be considered roughly stationary with respect to eye movements.

5.2.1. Slow Processes

Slow adaptation has been extensively studied since the beginnings of vision research (see Wyszecki & Stiles, 1967, for a review). The most widely studied effect is dark adaptation, the increase in sensitivity to light that follows complete darkening of the visual field. Smaller decrements of the mean luminance of the entire field and darkening of a portion of the field have received less attention.

Sensitivity increases for at least 10 minutes in full dark adaptation. The brightness of low-luminance objects increases as sensitivity increases. The effect of dark adaptation on surface-color constancy, on the other hand, will depend primarily on how it affects the accuracy of transfer of contrast from the retinal image. I know of no systematic study of lightness, apparent illumination, or apparent surface orientation as a function of dark adaptive state. Arend and Goldstein's achromatic Mondrian experiments show very clearly that lightness constancy does not require slow adaptation to the mean luminance of the stimulus array. Our subjects were able to accurately match reflectances in Mondrians more than one log unit less in mean luminance than the adaptation luminance (i.e., the mean luminance of the standard Mondrian). At the same time, the effects of mean luminance were not eliminated by adaptation. The brightnesses of the test and standard Mondrians were different when their mean luminances differed.

5.2.2. Fast Processes: Rapid Adaptation

In recent years there has been a great deal of work on changes of sensitivity over much shorter exposure times, < .5 s. Rapid adaptation is now widely thought to consist mainly of gain control in retinal processes (Shapley & Enroth-Cugell, 1984; Whittle, this volume, chap. 2; Whittle & Challands, 1969). This gain control allows retinal neurons with limited dynamic ranges to signal contrasts over a large range of mean luminances. Whittle and Challands (1969) showed that targets with equal local luminance contrasts produced equal brightness contrasts[10] over a wide range of high mean luminances. At lower mean luminances, however, it was necessary to increase the luminance contrast to maintain constant brightness contrast. Whittle and Challands' contours of constant brightness contrast closely resemble curves describing neural adaptation processes in mammalian retinas (Shapley & Enroth-Cugell, 1984). These data and similar data from other laboratories suggest that early visual processes succeed in approximately encoding local luminance contrasts at high mean luminances, but that they result in relative underestimation of luminance contrasts at low mean luminances. The underestimation occurs at mean luminances that are important in practical applications, luminances that frequently occur in both natural scenes and man-made images (television, photographs).

This loss of contrast efficiency at low luminances potentially affects surface-color perception. In my human intrinsics model the lightnesses of surfaces are derived from the output of early visual processes. Luminance contrast is proportional to the gradients of log luminance that are integrated in the Arend–Blake model (for contrasts below about 0.6; Arend, 1991). One might, therefore, speculate that the unasserted color image of the intrinsics model is derived from integrals of early visual signals encoding luminance contrast. If so, we might also expect that the reduction of apparent contrast at low luminances described by Whittle and Challands would affect lightnesses at low mean luminances.

There are two aspects of these hypotheses that are relatively new, the linking of neural contrast signals to spatial integration and the predicted consequences of loss of apparent contrast at low luminances.

Although there is widespread agreement that rapid adaptation allows the early visual system to encode luminance contrast (to the accuracy allowed by the biological processes), the role of the resulting information in brightness and lightness perception has been less clear. Shapley and

---

[10]For simplified reading I use *brightness contrast* to describe the criterion in Whittle and Challands' experiments, but this involves a simplifying assumption. In their paradigm, small test and comparison fields on circular background fields were presented to the right and left eyes, respectively, positioned so that the background fields were haploscopically fused. The test and comparison patches were in different locations, appearing side by side. Their subjects were asked to match the *brightnesses* of the test and comparison patches. The fused backgrounds had a common brightness, so matching the brightnesses of the two patches also made the differences between the brightnesses of the two patches and their common background nearly equal. By substituting the term *brightness contrast*, I am assuming that the physical contrast setting that makes the brightness of Whittle and Challands' test patch match that of the comparison patch would produce an approximate brightness-contrast match under normal viewing conditions, in which the backgrounds of the two patches are presented side by side rather than fused. Although this assumption has intuitive appeal, the normal-view experiments required to check it have not yet been done.

Enroth-Cugell (1984) argued that the contrast signals are sufficient to specify lightnesses, and Wallach's (1948, 1963) psychophysical writings seem to take the same position.

There are two problems with that view. First, contrast is a differential signal that indicates luminance ratios, not absolute luminances. The relationship between the lightnesses of two nonadjacent surfaces can only be determined from their local contrasts if they are known to be abutted by the same reflectance, but that is not generally true. As Land and McCann (1971) pointed out, all the contrasts over the region separating the targets must be integrated. Second, as discussed in section 3, contrast signals must be analyzed into their illumination and reflectance components, the "edge classification" issue described by Gilchrist et al. (1983).

Much of the confusion about the role of local luminance contrast in lightness perception can be traced to the disk–annulus patterns used in most earlier work: They are too simple for unambiguous interpretation in terms of lightnesses. The boundary between the disk and annulus can be equally easily perceived as a reflectance edge or an illumination edge (Fig. 4.7). In fact, it has been treated each way in various psychophysical

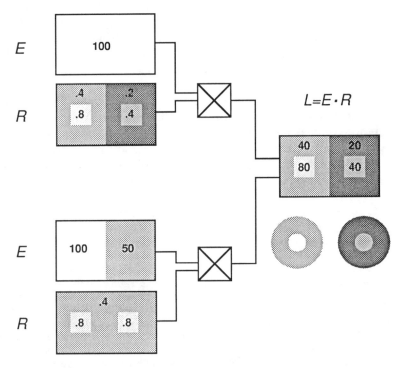

FIG. 4.7. Disk–annulus stimuli are perceptually ambiguous. Either of the physical arrangements on the left could produce the luminance patterns on the right.

experiments. Which physical situation the observers perceive depends on their instructions.

In our experiments with more complicated stimulus patterns ("Mondrians"), my collaborators and I have tried to reduce the ambiguity of the percepts. In one recent set of experiments (Arend & Spehar, 1993b), we were able to manipulate the stimulus such that changes in the luminance of an annulus surrounding the test patch were unambiguously perceived to be reflectance changes in one condition and illumination changes in another. The simplest explanation for the data is that subjects in experiments with isolated disk–annulus stimuli match brightness contrasts when they think they are matching lightnesses.

In other experiments (Arend & Spehar, 1993a) we obtained evidence that brightness contrast is influenced by early visual processes as expected from Whittle and Challands' work, but that apparent luminance (brightness) and apparent reflectance (lightness) are quite different functions of illuminance level. We measured lightness, brightness, and apparent contrast with disk-and-annulus patterns embedded in Mondrians (Fig. 4.8). In these patterns apparent reflectances are less ambiguous than in disk–annulus patterns in isolation. In all our previous experiments the reflectances surrounding the test and standard patches were identical. Now we placed test and standard disks on annuli of different grays. Under these conditions the subject can set the test patch according to each of three distinct criteria. Our subjects matched the apparent amounts of light coming from the patches (brightnesses) and their apparent reflectances (lightnesses), as in the earlier experiments. They also could match the brightness differences between the patches and their immediate backgrounds (brightness contrasts). With the stimuli of the earlier experiments with their identical reflectance contexts, lightness and brightness contrast tasks produced indistinguishable results, but with the different annulus lightnesses of the new experiments the three criteria produced quantitatively different results. The three patterns of the data are all readily referred back to the physical variables of the simulated scene.

The data from one subject are shown in Fig. 4.9. The illumination level on the test array (right side of Fig. 4.8) is indicated on the horizontal axis by the log of the ratio of the illuminance of the test array to the fixed illuminance of the standard array. The illuminations are equal at 0, and the test array illuminance decreases to the left. The subject adjusted the luminance of the test disk to make it match the standard disk, but it is more convenient for interpretation to plot the data as though the subject had adjusted the reflectance of the test disk. The subjects' mean log reflectance settings (mean log luminance − log illuminance) are plotted as ordinates. For comparison, the Munsell Values corresponding to the log reflectances of the left vertical axis are indicated on the right vertical axis.

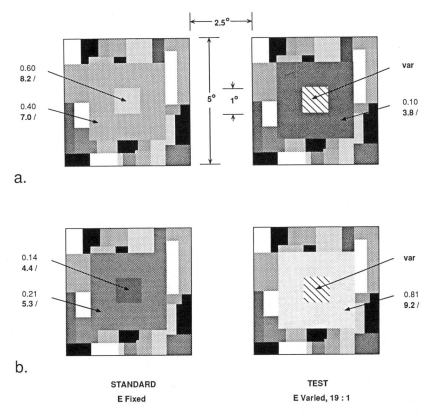

FIG. 4.8. Stimulus patterns from study of apparent contrast, lightness, and brightness (Arend & Spehar, 1993a).

The lightness and brightness data were similar to Arend and Goldstein's with one difference due to the different annulus reflectances. Lightness matches were illumination-invariant but were not quite reflectance matches; the different backgrounds of test and standard produced a small illumination-invariant error. This constant error was negligible for increments, but for decrements was approximately 1.5 Munsell Value steps (0.3 l.u.). Brightness matches varied substantially with illuminance. They lay between the reflectance-match and luminance-match lines.

To a first approximation brightness contrasts matched when the local disk–annulus luminance ratios of the test and standard were equal, but all three subjects showed a small but systematic deviation at low illuminances. Higher physical contrast was required at low mean luminances to match the standard (i.e., there was a loss of contrast efficiency).

The small deviations from luminance-ratio matches in Fig. 4.9 did not lead to failures of lightness constancy. To explore this further, I later

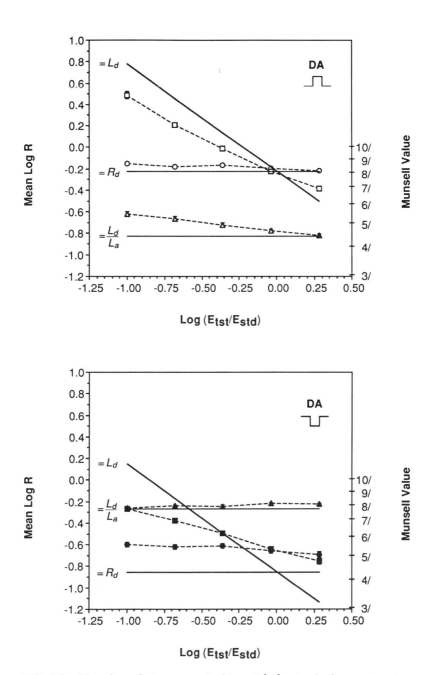

FIG. 4.9. Mean log reflectances required to match the standard array at various illuminances of the test array. Circles: Lightness matches. Squares: Brightness matches. Triangles: Brightness-contrast matches. Error bars are ± 1 s.e. ($n = 5$). If no bars visible, ± 1 s.e. is smaller than plot symbol.

extended the experiment to measure lightness, brightness, and apparent contrast at lower mean luminances (12.3, 1.23, and 0.123 cd/m^2). As anticipated from Whittle and Challands' data, much larger losses of contrast efficiency occurred at the lower luminances. To produce the required constant brightness difference between the test patch and its immediate surround, the observers required higher local luminance contrasts at low mean luminances. Lightness measurements, on the other hand, were affected by mean luminance at only the very lowest luminances. Surface-color perception was largely immune to the limitations of adaptation. One possible explanation lies in the relative nature of the lightness perception. The apparent contrast of the entire test Mondrian was lower at low mean luminances. As a result the test patch maintained its relative position within the perceived gray scale of the test Mondrian. Our subjects perceived the lightness of the test patch in the context of the reduced range of apparent contrasts in the test Mondrian. Gilchrist and Jacobsen (1983) also found that reducing the contrast of a scene did not always produce failures of lightness constancy.

This suggests that limitations of early visual processes need not always perturb perception of surface colors. On the other hand, simultaneous contrast does produce small errors of surface perception.

5.2.3. Simultaneous Contrast

Most introductory perception textbooks include a figure that demonstrates the phenomenon of simultaneous brightness contrast and shows that it can produce failures of lightness constancy. A gray patch is placed in the center of a white background, and a second patch with identical reflectance is placed in the center of the darkest black that can be printed. The patch on the white background is both a darker shade of gray and lower in brightness than the one on the black background. This failure of invariance with respect to changes of the background, a failure of lightness constancy, is usually attributed to retinal processes.

A number of authors have previously discussed background dependence as a failure of lightness constancy (Gilchrist et al., 1983; Koffka, 1931; Whittle, this volume, chap. 2; Whittle & Challands, 1969). The effect has been measured in several ways (Arend & Arend, in preparation; Takasaki, 1966; Whittle, 1991).

Takasaki (1966) used a pair of disk–annulus displays. He reported that equal steps on the Munsell value scale did not produce equal changes of lightness. One Value step above or below the background reflectance produced a much larger lightness change than did one Value step elsewhere in the reflectance scale. For example, with an annulus of Value 5/, changing the disk from Value 5/ to Value 6/ produced a much larger

4. SURFACE COLORS, ILLUMINATION, AND SURFACE GEOMETRY 189

lightness change than did changing the disk from Value 8/ to Value 9/. This exaggeration of steps near the background was called the *Crispening Effect*. Semmelroth (1971) published graphs to be used to adjust Munsell Values to compensate for the Crispening Effect.

As a study of apparent surface color Takasaki's experiment was flawed in two ways. As already discussed, disk–annulus stimuli are completely ambiguous with respect to surface interpretation. The luminance arrangement is just as likely to have resulted from an array of illuminances as reflectances and has in fact been used to represent both in experiments. Takasaki actually used papers to make his disks and annuli, but on the assumption that the apparatus was properly designed, it would not allow subjects to detect any differences between a construction from paper, a construction from projected light, or a CRT. If aspects of the stimulus other than the retinal illuminances of the test and inducing fields were important, it would be incorrect to characterize the stimulus and interpret the phenomena in terms only of the luminances of the disks and annuli.

Takasaki's paradigm also fails to make clear to the observer the distinction between lightness and brightness and which is to be reported.

To avoid those stimulus and response problems we have recently measured the influence of background reflectance on lightness and brightness (Arend & Arend, in preparation) using patterns similar to those of the mesopic lightness study. We placed disk–annulus patterns in the middle of Mondrians (Fig. 4.8). The Mondrian patches spanned the reflectance range from .03 (black) to .95 (white), providing enough context to allow confident perception of the position of the test patch in the gray scale, and the difference between brightness and lightness was carefully specified in instructions to the observers. The illumination on the standard Mondrian was fixed throughout the experiment, whereas that on the test Mondrian varied from trial to trial over a 19:1 range.

Several combinations of reflectances of disks and annuli were used. I describe only two here to illustrate the general pattern of the results. The standard disk was set at Munsell Value 3.0/, a slight increment when its annulus was set to 2.0/, and a slight decrement when its annulus was set to 4.0/. The test annulus was set to Munsell Value 9.8/. The subject adjusted the test disk to match the *lightness* of the standard disk. The two subjects' data were very similar; only one is described here. The mean reflectance settings for the described stimulus conditions are plotted as the bottom two data sets of Fig. 4.10. The standard disk was a slight increment for the condition plotted as open symbols and a slight decrement for the closed symbols. The solid line at the bottom of the arrow is the reflectance of the standard patch (i.e., the perfect lightness constancy line). The test patch had the same lightness as the standard patch when it was about 1.5/ Munsell Value steps higher in reflectance, consistent with expectations

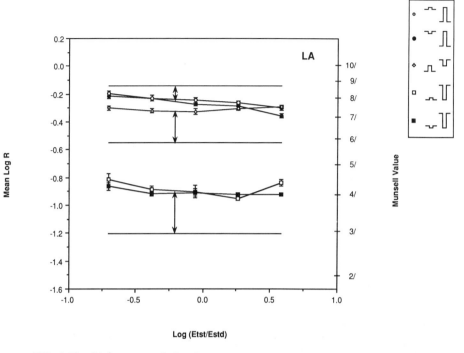

FIG. 4.10. Lightness match data from background dependence experiments. Mean log reflectance settings are plotted on the ordinate as a function of the log illuminance ratio on the abscissas. Right-hand ordinates are the Munsell Values that would produce the log reflectances on the left-hand ordinates.

from local luminance contrasts. The test patch on its white annulus needed a higher reflectance than the standard patch on its black annulus. The 1.5/ step error was independent of the illuminance of the test Mondrian. There was no evidence of the Crispening Effect in the lightness judgments; the difference between incremental and decremental standard patches is negligible. Using Semmelroth's graphs (Semmelroth, 1971), there should have been a difference of several Munsell Value steps between the increment and decrement matches.

Whittle's recent experiments (Whittle, 1992; Whittle, this volume, chap. 2) suggested an explanation for the discrepancy between my data and Takasaki's. He asked his subjects to adjust the luminances of 25 patches on a middle gray background so that they formed equally spaced lightness steps between black and white patches with fixed luminances. (A large range of disk luminances on a unified background is perceptually less ambiguous than isolated disk–annulus stimuli.) In a few sessions each patch was outlined by a thin black ring. In the other sessions the edge of each patch was simply formed by the patch luminance against the background luminance. Whittle found a substantial Crispening Effect, but only in the no-ring condition. In the ring condition there was no perturbation in the scale as the patches traversed the background luminance. The Crispening Effect was also greatly reduced when the disk and annulus had different chromaticities. The Crispening Effect occurs only when there is nothing but a simple luminance step delineating the disk, suggesting a link to the reduced luminance contrast at the edge. The luminance contrasts of the disks and annuli in my experiments were always large enough to be clearly visible. In everyday life we rarely encounter a surface that is completely surrounded by a very similar luminance of the same chromaticity. This suggests that the Crispening Effect plays little role in practical surface perception.

Whittle disagreed with this interpretation (Whittle, personal communication, August 27, 1992). He felt instead that the absence of Crispening Effect in my experiments was attributable to the complexity of my stimuli. The visual system may be able to discount local perturbations like the Crispening Effect when stimuli are complicated enough to support confident parsing of luminance gradients into illumination and reflectance gradients.

5.2.4. Threshold for Gradient

The threshold for spatial gradient in the Arend–Blake model and other gradient-integration models describes the well-documented insensitivity of the human visual system to shallow spatiotemporal gradients, another sensory limitation that may lead to errors of surface perception. Arend and Goldstein's (1987a) gradient illusions are all failures of correspondence between luminance gradients in the stimulus and perceived gradients.

As noted earlier, Land and others argued that this threshold makes a major contribution to lightness constancy. Our data on brightness and lightness in Mondrians under shallow gradient illumination (Arend & Goldstein, 1989) indicate otherwise. Even very small luminance gradients (compared to those common in natural scenes) were detected by our subjects and perceived as brightness gradients. Nevertheless, lightness matches in these scenes were illumination invariant. It seems likely that Land and McCann's (1971) unevenly illuminated-Mondrian demonstration is another example of the same phenomenon. A "black" patch at one edge of their Mondrian was given the same luminance as a "white" patch at the opposite edge by placing a light source closer to the former than the latter. The luminance gradients in their demonstration (a transparency bound in Land & McCann, 1971) are quite visible as brightness gradients under normal projection conditions. Nevertheless, the patches appear white and black, respectively, in agreement with our experiment. Furthermore, their gradients were much larger (0.8 log unit over the Mondrian)

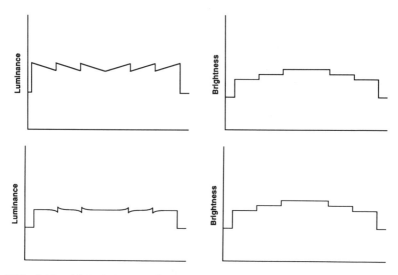

FIG. 4.11. (a) Radial sawtooth illusion (Arend & Goldstein, 1987a). (b) Craik–O'Brien–Cornsweet Illusion. When the shallow spatial gradients of the luminance distributions are very shallow, the brightness within each band is uniform, and the bands, identical in luminance profile, have different brightnesses.

than psychophysically determined thresholds for luminance gradients in otherwise unstructured fields (0.08 l.u./10 deg vis angle; (Campbell, Johnstone, & Ross, 1981; van der Wildt, Keemink, & van den Brink, 1976). There is no doubt that the threshold for spatial luminance gradients suppresses visibility of some very shallow illumination gradients in natural viewing. More generally, however, illumination gradients can be above threshold, with no impairment of lightness constancy.

Rather than assisting lightness constancy, the threshold for gradient may lead to errors under some conditions. Some shallow gradient patterns seem to produce distortions of both brightness and lightness. The radial sawtooth illusion (Arend & Goldstein, 1987a) and Craik–O'Brien–Cornsweet rings (Arend et al., 1971) are examples, when the luminance gradients are shallow enough that the bands appear perfectly uniform (Fig. 4.11). At higher contrasts the radial sawtooth illusion produces visible brightness gradients that often look like shading on a slanted surface (see also Bergström & Putaansuu, 1988).

5.2.5. Induced-Brightness-Sawtooth Illusion

The induced-brightness-sawtooth illusion (Arend & Goldstein, 1987a) is another example of an illusory brightness gradient producing a perceptual error, the appearance of depth. In this case it is a consequence of inconsistencies in the 2D pattern of gradients that remains after thresholding.

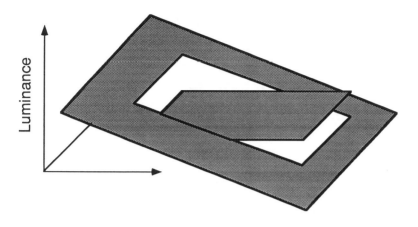

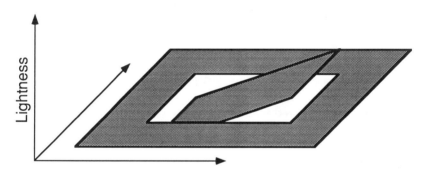

FIG. 4.12. The induced-brightness-sawtooth illusion (Arend & Goldstein, 1987a). Viewed from a close distance there is a gradient of brightness from left to right in the central rectangle and the surrounding area appears fairly uniform in brightness (on a large CRT display it can be very uniform in appearance). It is, however, the surrounding area that has a spatial luminance gradient—the luminance of the central rectangle is uniform.

We surrounded a rectangle of constant luminance with an annular field across which there was a shallow luminance gradient from left to right. If the surround gradient is sufficiently shallow, there is no brightness gradient in the annulus. Instead, an illusory brightness gradient of opposite sign is perceived in the central rectangle. There is usually a strong impression that the rectangle is tilted out of the plane of the display, and the induced brightness gradient is perceptually attributed to surface orientation rather than reflectance.

The location of the brightness gradient in this illusion has strong implications for the integration algorithms of gradient-integration models. The visual threshold for luminance gradients makes the pattern of detected spatial luminance gradients in this stimulus inconsistent. Luminance gradients across the top and bottom edges of the rectangle signal that there is a changing luminance difference in this direction, but there are no suprathreshold gradients in the horizontal direction anywhere within the rectangle or surround. Inconsistency in the pattern of gradients from the various edges requires that models predict a brightness gradient somewhere. The location of the perceived gradient is determined by the model's integration algorithm.

No current model predicts the actual location of the brightness gradients in this illusion. Relaxation integration algorithms like those used by Horn (1974) and Blake (1985) distribute the inconsistencies evenly over the entire field, predicting brightness gradients in both the annulus and rectangle. Local contrast and filling-in mechanisms (Davidson & Whiteside, 1971; Fry, 1948; Grossberg & Todorovic, 1988) also predict gradients in both the rectangle and annulus. Another possible integration algorithm, an adaptive threshold adjustment algorithm (Arend & Goldstein, 1987a), produces brightness gradients only at locations where there are luminance gradients in the input image. We rejected it on the basis of this illusion.

The illusion also has a chromatic counterpart. An orange rectangle of spatially uniform chromaticity is surrounded by an annulus with a shallow gradient of chromaticity such that the yellow component increases and the red component decreases from left to right. As in the brightness case, a gradient of opposite sign is perceived in the central rectangle; the orange gets gradually redder and less yellow from left to right.

5.3. Psychophysics and Intrinsic Images

One of the stated goals of this intrinsic image model is facilitation of testing of algorithms against human experimental data. Whereas the model is useful in making the difficulties more explicit, it needs further elaboration to allow confident empirical tests. Some of the difficulties are not easily solved.

5.3.1. Difficulties in Measuring Human Intrinsics

The entries in the arrays of the model in Fig. 4.5 are internal computational variables. In machine vision models they are directly known and explicitly linked to machine vision performance measures. In human models, however, the variables represented in conscious perception are not necessarily the actual computational variables. Logically, they are related by some unknown, possibly complicated, transformation. To test hypotheses about computational algorithms we have only psychophysical measurements of the perceptual variables, data even further removed logically from the computational variables.

There is, for example, no way to directly measure variables at the level of the input to the intrinsic images computation. Psychophysical responses, whether verbal or motor, occur at the end of a chain of processing that includes at least surface computations, object computations, decision processes, memory, and linguistic manipulations.

There are at least two ways we might describe our simultaneous perception of surface properties and properties of objects computed by more central modules. In a *parallel access* metaphor (Fig. 4.13a) the central structures receive not only the output of object computations but also copies of the intermediate images computed by earlier stages. The *serial* metaphor (Fig. 4.13b) is a kind of modified top-down model. In this view the central structures have direct access to only the output of the object computations. Perception of surface properties and unasserted colors are indirect via elaborate computations. Surface properties are derived from the object percept by approximately inverting the computations lying between the intrinsic images and the object representation. Unasserted colors require inversion of both the object computations and the surface computations.

These two models have different implications with respect to the impediments to testing of hypotheses about human surface and object computations. In the serial case one must contend not only with possible transformation between the computational variables and the dimensions of the conscious representation but also with the possibility of inaccurate inversion of the object and surface computations.

This picture may make the prospects for systematic study seem bleak, but the situation is logically no different from traditional psychophysics. When we view general, unrestricted scenes, the surface and object computational algorithms are provided with rich input information. Reducing the stimulus to a single spot of light in a dark surround or to a disk–annulus denies the higher order processes the data they need for normal computations, placing them into an abnormal state. However, this does not necessarily cause them to pass sensory signals to higher levels unaltered. Arguments for a causal relationship between peripheral physiological responses and psychophysical data are based on close correspondence between the data patterns from the two domains. It is always possible that higher level processes make substantial contributions to the form of the psychophysical data. Although few psychophysical problems can be considered completely solved, the enterprise has at least produced a large body of reasonably orderly data.

To generate testable hypotheses about surface perception it is necessary to have a detailed model specifying not only the variables to be measured and their role in the computational chain but also hypotheses linking the theoretical variables to measurable quantities. For example, consider a

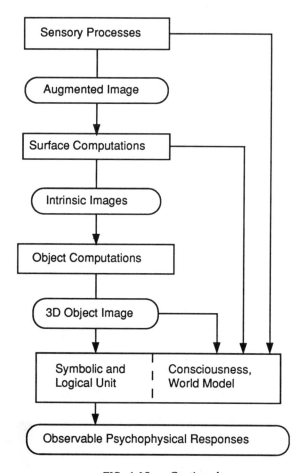

FIG. 4.13a. Continued

simple possible relation between the output of Stage 1 and measurable quantities:

> Linking Hypothesis: The unasserted color array is a direct perceptual representation of the input array of the intrinsic-image computation. The Stage 2 constancy computation has no access to retinal image information not represented in this array's hue, saturation, and brightness.

By this hypothesis, stimulus regions that psychophysically match in brightness provide the same input level to the intrinsics computation. If gains are set by early adaptation mechanisms, the intrinsics computation knows only their output, not the gains. With respect to simultaneous contrast, the

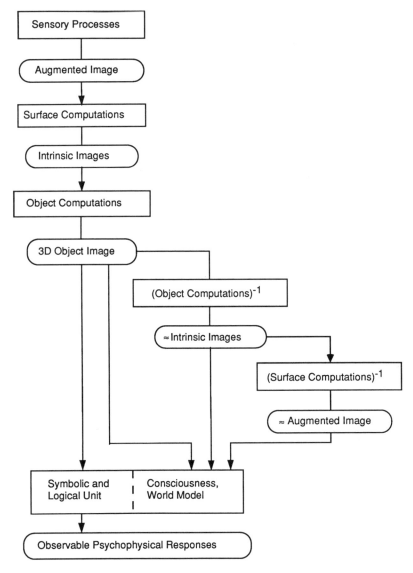

FIG. 4.13. (a) Diagram of parallel access ("bypass") metaphor of subjective spatiotopic representation. (b) Diagram of serial access ("undoing") metaphor of subjective spatiotopic representation.

computation has access to only the response as modified by interaction with neighboring regions. If early visual processes do not perfectly transfer relative luminances, the errors may propagate into the intrinsics computations.

Let us now try to interpret the Arend and Goldstein (1987b) achromatic constancy data. The constancy of the lightness data shows that our "paper-matching" task provides an operational definition that corresponds well with previous perceptual definitions of lightness as perceived reflectance. Under the conditions of our experiment, papers match in lightness when their reflectances are equal.

Because lightness matches and perceived geometry were invariant over illuminant change, Equation 3 indicates that the remaining variables, brightness and perceived illuminance, should be proportionally related. The shallow negative slopes of the brightness-match data (constant-brightness curves) imply shallow positive slopes for perceived illuminance in log–log coordinates. We did not attempt to independently measure perceived illuminance,[11] but there is reassuring consistency within the lightness and brightness-match data. For reliable perception, perceived illumination should be independent of the reflectance being matched. The brightness matches are not luminance matches, but the slopes of the brightness curves are nearly equal for the four Mondrian curves. Because the lightness curves are independent of illuminance, the equal brightness slopes suggest that the perceived illumination was the same for the four patch reflectances. In the disk–annulus case the situation is more complicated, but they are consistent with the Mondrian condition in the following way. The brightness curves are not parallel, converging by approximately 0.4 l.u. over the 1.28 l.u. illuminance range. This does not, however, necessarily mean that the perceived illuminances changed with the reflectance of the standard patch. The lightness curves converge by approximately the same degree as the brightness curves. As a consequence $\log(R_{brightness}/R_{lightness})$ is roughly independent of patch reflectance.

5.3.2. Psychophysical Measurements of Intrinsic Images

Theoretical developments in image analysis have inspired a number of recent efforts to measure intrinsic images of human subjects. Todd and his colleagues (Mingolla, 1983; Mingolla & Todd, 1986; Todd & Mingolla, 1983) measured apparent surface orientation in computer-generated shad-

---

[11] It is difficult to design an illumination measurement task for the laboratory that is convincingly independent of other perceptual tasks. For example, if two patches are perceived by the subject to have the same reflectance, the subject can adopt a strategy of setting the variable illumination to make the equal-reflectance patches have equal brightnesses. The resulting data might not therefore be independent of the data from lightness and brightness-matching tasks.

ing patterns. Bülthoff and Mallott (1987, 1988) have investigated the combined influences of stereopsis and shading on apparent surface orientation. Blake and Bülthoff (1990) showed that subjects' perception of surface orientation was influenced by the stereoscopic depth of a highlight, indicating that psychological physics encompasses fairly sophisticated geometrical optics.

I avoided mentioning transparency until now for fear of overwhelming the reader with complexities. The problem of computing surface properties described earlier, already complicated, has been greatly simplified by my tacit assumption that all surfaces are opaque, that there is only one surface visible along each line of sight. In practice, of course, we occasionally encounter scenes that include transparent surfaces. The transparent layer has its own set of intrinsics (see Gerbino, 1988) in addition to those of the opaque layer. The optics of achromatic transparency are already complicated, including additive components as well as multiplicative, but many of the quantities are wavelength dependent as well. Gerbino et al. (1990) found that subjects' perception of transparency in simple achromatic patterns was predicted by a model expressing the mathematics of transparency in such patterns. The fact that we are generally quite competent in perceiving transparency underscores the elegance of our visual systems' perceptual computations and makes all too clear the scope of the task ahead.

6. CHROMATIC SURFACE PERCEPTION

It is not possible in this chapter to fully generalize the preceding arguments to include chromatic surface-color perception, but it is important to briefly illustrate some of the issues. Most of the achromatic problems become much more complicated when we consider chromatic responses to the world. The customary description of the problem of inverting the achromatic photometric equation (3) is a gross oversimplification of the real problem. Even retinal illuminance, the usual description of a physical achromatic image, is based on the action spectrum of a complicated joint response of the four receptor classes.

6.1. Chromatic Variation of Illumination Within Scenes

I argued earlier that univalued algorithms are inadequate as constancy mechanisms because the local illumination varies within most natural scenes. The arguments also apply in the case of chromatic surfaces and lights, for the same reason: The color of the illumination varies from point to point within a single scene.

Shadows and shading strongly affect the distribution of achromatic illuminance, but it is not so immediately obvious that they cause important *chromatic* illumination gradients in three-dimensional scenes. If a scene is illuminated by only a single small source, the chromaticity of the local illumination on a Lambertian surface depends much less on surface orientation than does its illuminance. Several image analysis algorithms (Klinker, 1988; Nikolaev, 1988a, 1988b; Shafer, 1985) have attempted to exploit this to separate illumination from reflectance.

Unfortunately, 3D scenes with a single, small light source are rare in our environment due to illumination by light reflected from other surfaces, and we therefore cannot rely on mechanisms that need a single illumination color to produce surface-color constancy. Shading and shadows *do* produce important gradients of illuminant chromaticity in the usual case, scenes with two or more spatially separated, spatially extended, different-colored sources of illumination. At most points in natural settings light reflected from nearby colored surfaces serves as a secondary illumination, with strength that is often an appreciable fraction of the direct light from the sun or artificial light sources (Forsyth & Zisserman, 1990). For example, consider an office with a glass wall, surrounded by other buildings. All illumination at points from which one cannot see the sky (most of the room) is by secondary reflection. This secondary illumination is strongly chromatic near brightly colored surfaces. Its effect depends on the relative orientation of the illuminating and illuminated surfaces. As the illuminance from one source decreases due to shading or shadows, the proportion of the total illumination from other sources is increased. Consequently, the color of the effective illuminant can vary dramatically from point to point within a scene.

Even the variations of daylight color within a scene are important. An object facing upward in an open meadow an hour after sunrise will be illuminated by both reddish light (about 4000 K) from the sun's direction and bluish light (about 10000 K) from the rest of the sky (Wyszecki & Stiles, 1967). Portions of the surface of an object with different orientations are illuminated by different mixtures of these lights. To illustrate the effect of this, the chromaticities of Arend and Reeves' 32 Munsell papers under three daylights are plotted in Fig. 4.14. The effects of daylight variation depend on the spectral reflectance of the surface. The arrow points to a greenish-blue paper that has the same chromaticity under the reddish illuminant (4000 K, triangle) as a yellow paper has under the bluish illuminant (10000 K, square). Maximov (1984) reported a striking related observation. When in a shadow the petal of a dandelion is illuminated by north skylight and consequently has the same chromaticity as grass in direct sunlight. Nevertheless, the petal appears yellow and the grass green.

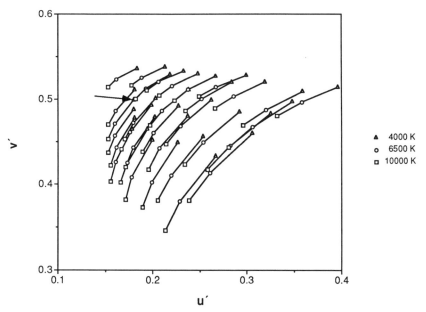

FIG. 4.14. Chromaticities of 32 Munsell papers under 4000 K (triangles), 6500 K (circles), and 10000 K (squares) daylight illuminants. Arrow indicates a greenish-blue paper under 4000 K that has the same chromaticity as a greenish-yellow paper under 10000 K.

6.2. Failure of Single-Dimensional Models: Color Appearance

Because the local illuminant color varies substantially within scenes, my arguments regarding failure of achromatic sensory processes (e.g., adaptation and contrast) as lightness constancy mechanisms apply equally to chromatic sensory processes as surface-color constancy mechanisms. The suprathreshold effects of chromatic mechanisms have been extensively studied under the rubric *color appearance* (see Wyszecki, 1986, and Pokorny, Shevell, & Smith, 1991, for reviews).

Consider, for example, chromatic adaptation. If we change the color of the illumination falling on a gray surface, prolonged chromatic adaptation shifts unasserted colors in approximately the direction required to compensate for the illuminant change (Arend, 1993a; Helson, Judd, & Warren, 1952; McCann, McKee, & Taylor, 1976; Worthey, 1985). The unasserted color (the color of the light coming from the surface) after adaptation will approximate that under the previous illuminant. The approximation is, however, not very good. In recent experiments (Arend, 1993a) designed to maximize chromatic adaptation and color contrast (long adaptations, near-white illuminants), the adaptive shifts were only about 65% of those needed to produce illuminant-invariant unasserted colors.

A number of researchers have explicitly considered this approximate invariance of the color of the light to be equivalent to constancy of apparent surface color (e.g., Dannemiller, 1989; Hallett, Jepson, & Gershon, 1988), and many others have implicitly assumed this equivalence by failing to make a clear distinction between unasserted color and apparent surface color (e.g., Brainard & Wandell, 1986; D'Zmura & Lennie, 1986; Helson et al., 1952; McCann et al., 1976; Valberg & Lange-Malecki, 1990; Worthey, 1985).

Our experiments (Arend & Reeves, 1986; Arend et al., 1991) showed that prolonged adaptation and its attendant alteration of unasserted colors are not necessary for approximate constancy of surface colors. The patches in the test Mondrians under 4000 K and 10000 K illuminations were viewed with adaptation approximating that appropriate for the standard Mondrian illuminant (6500 K), and their measured unasserted colors were different from those of their counterparts in the standard Mondrian. The data showed little adaptation of unasserted colors during the 1-second fixations we allowed the subjects. Nevertheless, the apparent surface colors in the test Mondrian were approximately the same as those in the standard Mondrian. The time course of chromatic adaptation is discussed further in section 6.4.

I know of no other chromatic experiments requiring observers to distinguish judgments of unasserted color from apparent surface color. However, Judd and Wyszecki (1975) reported a related observation:

> When we pass from natural daylight into a room illuminated by incandescent-lamp light we notice immediately that the color of the light reflected from the objects in the room has changed. Objects that reflect green light in daylight now appear to reflect yellow-green light; if the object reflects purple light in daylight, it now appears to reflect a more reddish kind of light. *This immediate change in perceived color, however, does not appear to apply to the object.* That is, we recognize that the color of the light is reddish-yellow and perceive the object color relative to this, that is green much as it was in daylight. (Italics added, p. 354)

In other words, constancy of apparent surface colors occurs in *both* appropriate and inappropriate adaptive states, despite the dramatic change of unasserted colors.

As in the achromatic case, chromatic sensory processes have no way of distinguishing spatial gradients of illumination from spatial gradients of reflectance. As a consequence they cannot produce illumination-invariant surface appearance in multiple illuminant scenes. They are important, but they are not by themselves perceptual constancy mechanisms.

6.3. Chromatic Intrinsic-Image Model

Thus, in the chromatic domain, too, sensory processes can only serve to condition information before other processes compute perceptual interpretations in terms of surface colors, colored illuminants, and scene-domain geometry.

In the model (Fig. 4.5) each of the perceived achromatic intrinsic images has a chromatic counterpart. Generalization of the achromatic model to include chromatic stimuli and responses introduces several new problems.

The chromatic information available to visual computations is strictly limited by the photoreceptors' sampling of the spectrum (i.e., the number of receptor types with independent spectral sensitivities). In human vision the four receptor systems (rods, three cone types) can be thought of as providing four independent images of the scene, in different wavebands. The color information available for computation of surface properties is that from these four images after they have been transformed and combined by sensory processes.

The situation is also more complicated at the level of the intrinsic images. Each of the single-valued images of the achromatic model is replaced by a multidimensional color image. For example, perceived surface colors vary in lightness, hue, and saturation over the spatial array.

6.3.1. Between-Intrinsic-Image Constraints: Computational Color Models

In the general case, the achromatic image formation equation, $L = E R \cos \phi$, is replaced by four equations with integrals over the spectrum. The signal (i.e., the receptor excitations) generated by the light at a point in the retinal image is described by

$$\rho_\kappa = \int \Phi(\lambda) E'(\lambda) S(\lambda) R_k(\lambda) d\lambda \qquad k = 1, \ldots, 4 \qquad (4)$$

where $E'(\lambda)$ is the illuminant power spectrum, $S(\lambda)$ is the reflectance spectrum of the viewed surface, $\Phi(\lambda)$ represents the photometric geometry of the surface, eye, and illuminant, and $R_k(\lambda)$ is the spectral sensitivity of receptor system k. As in the achromatic case it is convenient to consider $\Phi(\lambda)$ as affecting the effective illuminant power so that the equation reduces to

$$\rho_\kappa = \int E(\lambda) S(\lambda) R_k(\lambda) d\lambda \qquad k = 1 \ldots 4 \qquad (5)$$

where $E'(\lambda)$, the local illuminant power distribution, is replaced by $E(\lambda)$, the effective illuminant power.

These integrands, in principle, have infinite degrees of freedom, corresponding to the continuum of wavelengths. In an unconstrained world each coefficient in the integrand of Equation 4 could take on any of an infinitude of values at every wavelength. The visual system would be confronted with the achromatic inverse optics problem at each wavelength. Because human photoreceptors provide only four independent spectral samples, the computational problem would be hopelessly underconstrained, and even crude surface-color constancy would be impossible.

However, our environment is not random; it is constrained by optical physics. Common surfaces and light sources have highly redundant spectral distributions. Because reflectance varies gradually over the spectrum, the reflectance at one wavelength is highly predictable from the reflectances at neighboring wavelengths. This redundancy means that the variations among spectral distributions of surfaces and lights can be captured by the responses of a relatively small number of well-chosen sensors. Part of our visual systems' knowledge of that physical structure is in the form of the four independent receptor systems that have evolved to sample the small piece of the electromagnetic spectrum known as visible light.

Several theorists (Brill, 1978, 1979; Buchsbaum, 1980; Maloney, 1985; Sällström, 1973) have analyzed the implications of this informational constraint for color constancy. They have proposed algorithms for computing surface colors based on *finite-dimensional* computational models. Finite-dimensional models of color computation describe linear algebraic relationships among the input image and the intrinsic images that are imposed by optical physics, given the models' various assumptions. One interesting accomplishment of this theoretical work is that it has progressed without requiring significant contributions from experimental measurements of human surface perception. The negative side of their abstract nature is that it is hard to compare their formal properties to those of human surface perception.

I do not try to give a thorough analysis of these models here. However, a few comments about their implications for chromatic intrinsic-image models may be useful. First of all, the finite dimensional models provide several benefits regardless of their success or failure as computational algorithms. They provide relatively simple equations describing the relationship among the intrinsic images, analogous to the achromatic image formation equation (1). They also provide expressions characterizing the values in the intrinsic images. It was attempts to design experiments testing these models that first convinced me of the need for a heuristic intrinsic-image model. The model sketched in Fig. 4.5 evolved from my attempts to make explicit the concepts and relationships necessary for experimental evaluation of color constancy algorithms.

4. SURFACE COLORS, ILLUMINATION, AND SURFACE GEOMETRY   205

Second, they clarify the conditions required if one is to compute a unique solution for surface colors solely on the basis of the set of excitations of the visual system's receptors. Natural scenes frequently violate the required conditions. Our (at least) crude competence at surface-color perception implies that the human visual system uses information beyond that represented in these models.

For example, consider Maloney's (1985) exceptionally clear analysis. He showed by means of linear algebraic arguments that a visual system with p classes of photoreceptors with independent spectral sensitivities can recover illumination-invariant descriptors of surface colors (i.e., have surface-color constancy), given a set of well-defined restrictions on the optical physics of the world being viewed. A single illuminant with p or fewer degrees of freedom must fall on at least $p-1$ independent reflectances from a population of reflectances that varies along $p-1$ or fewer dimensions. With three receptor classes, for example, the data from two or more reflectances under a single illuminant define a plane in the three-dimensional response space of the receptors (Fig. 4.15), provided that the reflectances are at most bivariate and the illuminant at most trivariate.[12] In a world with more degrees of freedom for either reflectances or lights, the algebra allows no unique solution for surface colors. If the illuminant spectral power distribution and surface reflectance distribution are linear combinations of fixed basis functions and if the visual system knows those basis functions and the action spectra of its own photopigments, then the illuminant and reflectance spectral distributions can be approximately recovered, to a multiplicative constant, from this plane.

Before turning to problems with meeting the assumptions, it is important to be clear about the degree of constancy achieved when they are met. The ambiguity of the multiplicative constant (common to all these models) means that the solution completely specifies only the chromatic dimensions of the surfaces and lights. Figure 4.16, an elaboration of Fig. 4.15, may help make the implications clearer. The excitations are all positive in this more realistic model, and the lights from the scene lie in the planar wedge defined by the origin and the chromaticities of the two surface basis functions under the illuminant. Responses lying along any particular vector direction, α, β, in Fig. 4.16 all have the same ratios of receptor excitations (i.e., they have the same chromaticity). The solution gives the

---

[12]Although reflectances and illuminants in our natural environment have relatively smooth spectra, they are probably not well described by so few dimensions. Possible input from a fourth receptor system, the rods, is not considered here, and the other models include algorithms that yield an additional degree of freedom in the surfaces (if their additional assumptions can be met). The three-receptor example is used here because it is easier to visualize than higher dimensional data.

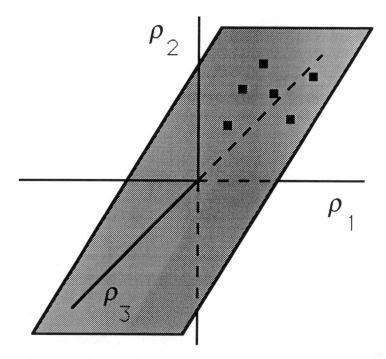

FIG. 4.15. Solution plane for a single illumination in a three-receptor example of Maloney (1985) algorithm. The axes are levels of excitation of three types of photoreceptor. The points are the excitations produced by six colored surfaces under a single illuminant. The shaded plane is the loci of all spectral reflectances under that illuminant, given the Maloney assumptions about the properties of photoreceptors, reflectances, and illuminants.

orientation of the plane and the radial direction of the surface within the plane, so the chromaticity that the surface would have under any chosen standard white light is completely known.

The vector length, a, represents the amount of light absorbed by the receptors. There are two problems for lightness constancy regarding these vector lengths. Any given length may be due to either intense light falling on a less reflective surface or less light (with the same relative power spectrum) falling on a lighter surface (with the same relative reflectance spectrum). On the other hand, if we increase the illuminance, all the vectors corresponding to various surfaces will increase proportionally. Thus the solution provides relative luminous reflectances only. This first problem might not be a major practical limitation, especially if the surfaces approximately span the reflectance factor axis of the color solid (but it does require further computation for lightness constancy). The second limitation is more complicated. If one substitutes for the illuminant of Fig. 4.16 an illuminant with a different spectral power distribution but the same illuminance, the vector lengths for various surfaces will change disproportionately, depending on the correlations of their reflectance spectra with the illuminant SPD. For example, a red surface will have a higher luminance relative to a white surface as the illuminant becomes redder, and a blue surface will have a lower reflectance factor. For spectrally selective surfaces under spectrally selective lights, lightness constancy requires that the color of the illuminant be taken into account. Given that the color of the illuminant is part of the solution and that the model applies to only fairly smooth reflectance spectra, it may be possible to formulate a function of vector angle in the solution plane that would null this effect on reflectance factor. Although lightness constancy for chromatic surfaces and illuminants has been studied much less than achromatic surfaces, the evidence sug-

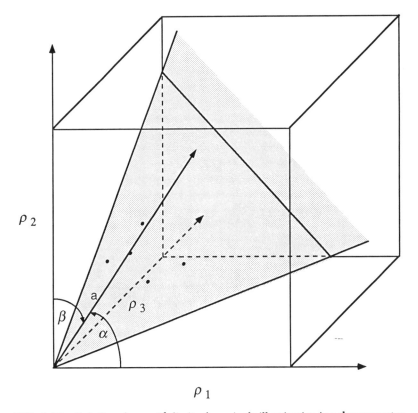

FIG. 4.16. Solution plane, with limits, for a single illumination in a three-receptor example of Maloney (1985) algorithm. The axes are levels of excitation of three types of photoreceptor. The points are the excitations produced by five colored surfaces under a single illuminant. The planar wedge is the loci of all spectral reflectances under that illuminant, given the Maloney assumptions about the properties of photoreceptors, reflectances, and illuminants.

gests that good lightness constancy can obtain (see, e.g., Sobagaki, Yamanaka, Takahama, & Nayatani, 1974).

Before we turn to human surface perception, consider Maloney's model as an abstract image processing system. The visual system described by the model produces a solution by supplementing the signals from its sensors with a number of constraints, assumptions that it makes about the world it is viewing. To begin with, it imposes constraints of the between-intrinsic-image variety by assuming that the imaging physics obeys Equation 4. To reduce the dimensionality of the problem it also makes several specific within-intrinsic-image assumptions. It assumes that the chromatic reflectances, $S(\lambda)$, and illuminant distributions, $E(\lambda)$, can be represented as a linear combination of several (in our example, $n = 2$ and $m = 3$, respectively) basis functions, S_j and E_i with weights σ_j and ε_i, respectively:

$$\rho_\kappa = \int E(\lambda) S(\lambda) R(\lambda) d\lambda$$

$$= \int \left[\sum_{i=1}^{n} \varepsilon_i E_i(\lambda) \right] \left[\sum_{j=1}^{m} \sigma_j S_j(\lambda) \right] R_k(\lambda) d\lambda$$

$$= \sum_{i=1}^{m} \sum_{j=1}^{n} \varepsilon_i \sigma_j g_{ijk} \qquad (6)$$

where $g_{ijk} = \int E_i(\lambda) S_j(\lambda) R_k(\lambda) d\lambda$.

The g_{ijk} are constants; they include only fixed, scene-independent quantities that the visual system is assumed to know.

These assumptions, including a particular fixed set of basis functions and photoreceptor action spectra, are part of the visual system's psychological physics, its model of the world. If the system tries to analyze images from a world with optical physics different from those assumed, it will produce no solution or incorrect solutions. This might occur, for example, because imaging in this alternative world does not obey Equation 4, or because the reflectances and/or illuminations are outside the gamuts (spans) of the basis functions (either because the basis functions are the wrong ones or because the dimensionality of the reflectances and illuminations are too high).

Now let us consider how finite-dimensional color models are to be related to human vision. Maloney's algorithm is designed to receive as input the signals from p receptor systems and find a set of n numbers describing reflectances and m numbers describing illuminants. These are ε_i and σ_j, the weights on the basis functions in the linear sums specifying the apparent reflectance and apparent illuminant spectra (Equation 5). Humans probably do not need to know the actual spectra. It is sufficient to have the σ_j, which are illuminant-invariant descriptors of surface color. They might be loosely thought of as the appearance of the surface if it were under some canonical illumination (e.g., an equal energy white light).

In machine vision research all the quantities in the model are directly known. Testing the algorithm consists of comparing the computed surface and light descriptors to those directly measured in the input scene.

Human vision models have an additional requirement. For the model to be testable we must be able to specify the mapping of the surface descriptors in the internal computation (left column, Fig. 4.5) onto psychophysically measurable dimensions of surface perception (right column, Fig. 4.5). However, research involving a clear distinction between measurements of apparent surface color and measurements of unasserted colors has only just begun (Arend & Reeves, 1986; Arend et al., 1991). The methods are still controversial. Convincing tests of computational algorithms as candidate human vision models will probably have to wait for further development and widespread acceptance of new measurement paradigms.

Some of the response variables of the intrinsic images are fairly familiar and have been successfully measured in previous research. For example, unasserted colors are well described in terms of the hue, saturation, and brightness dimensions. The dimensions of surface color have been characterized by the axes of several color order systems. For example, the Munsell system, under its prescribed viewing conditions, produces apparent surface colors that can be organized as Hue, Chroma, and Value dimensions. On the other hand, the perceptual dimensions of apparent illumination and apparent geometry have not been investigated to the same extent.

6.3.2. Within-Intrinsic-Image Constraints

There are a number of important issues in chromatic surface perception that fall under the rubric "within-intrinsic-image constraints." The representation of surfaces and lights in terms of known basis functions utilizes one form of within-intrinsic-image constraint, the slow variation over wavelength of the natural surfaces and lights. There are also other ways that physical structure in chromatic reflectances might be exploited by the human visual system. Finally, some of the within-image structure of the natural environment poses extremely serious problems for color constancy algorithms.

Range constraints might be applied within the surface-color images. The visual system can reliably deduce that a region of the input image contains an illumination gradient if the range of luminances and chromaticities exceeds the gamut realizable with a single illuminant and real reflectances. For example, in my example of the Maloney algorithm the full gamut of chromaticities realizable by reflectances under a single illuminant lies in the solution plane. Chromaticities outside the plane exceed the gamut and imply the presence of at least one additional illumination in the scene, with its own plane in receptor excitation space.

Even within a single-illuminant plane there are gamut constraints. Physically realizable colors lie in a small region of the solution plane. Reflectances are required to be between zero and one in all spectral bands, limiting surface colors to the gamut bounded by optimal colors (MacAdam, 1935). Forsyth (1988) developed a computational algorithm that added this constraint to a finite dimensional linear model of the Sällstrom variety and reported that it provided a very strong restriction of the solution space.

On the other hand, there is evidence that the human visual system uses a much more complicated algorithm than simple range constraints. In Arend and Reeves' (1986) Mondrian displays none of the patches under any of the three illuminants, 4000 K, 6500 K, and 10000 K, were outside the gamut of Munsell papers illuminated by 6500 K. Thus the test and standard Mondrians were always physically consistent with perception of two different reflectance arrays under the same illumination. Nevertheless, the pairs of Mondrians (6500 K vs. 10000 K, 6500 K vs. 4000 K) appeared to be two identical reflectance arrays under two different illuminations. When the 6500 K Mondrian is abutted to the 10000 K Mondrian, the receptor excitations are physically consistent with interpretation as a single rectangular array of 64 papers under 6500 K illumination, but instead the display appears to have a sharp illumination edge in the middle, separating two different illuminants.

Systematic data on environmental color is very scarce, but our Mondrian situation is probably in accord with ecological reality. The spectral reflectance gamut of many outdoor scenes is relatively small. If we exclude flowers and fruits, most reflectances are a small subset of the Munsell gamut, reds, yellows, and greens of various purities, and scenes with even smaller subsets are not rare. But we argued in section 6.1 that most natural scenes include spatial gradients of illumination color. It is therefore likely that the range of receptor excitations in many natural scenes under various daylights will not exceed the gamut of a slightly broader set of physically possible reflectances under a single illuminant.

The computational color algorithms posed to date fail completely when confronted with this overlapping gamut problem. All the sensory evidence lies within the intersections of the gamuts of multiple illuminations, each of which implies a different set of surface colors. They therefore lie in a single solution subspace (e.g., the plane in Fig. 4.15) of the receptor excitation space, but it is common to several illuminations. Even worse, in cases like the Arend and Reeves Mondrians the two illuminant sets are not even segregated within the solution subspace. The sets overlap over the majority of their range, as can be seen in the chromaticity diagram of Fig. 4.14. The algorithms cannot explain why such patterns appear to the human observer to be a smaller range of reflectances under multiple illuminants.

Thus far the authors of finite dimensional models have analyzed only input scenes having a single illuminant. If a scene with two illuminants contains reflectances from the nonoverlapping portions of their gamuts, the three-receptor system of Fig. 4.15 produces points that define two planes, with perfectly confounded points along the intersection. Even with an unusually broad range of reflectances, the frequently encountered case of a smooth gradient of illumination will produce receptor excitations with very difficult geometry. Each illumination along the gradient produces points from a different plane, generating a three-dimensional cloud of points.

It is possible that the existent models could cope with such scenes if extended to take advantage of spatial proximity and continuity constraints. Maloney (1985) briefly speculated about how his analysis could be extended to illumination gradients. Klinker (1988) has begun to deal with more complex scenes. None of the algorithms to date succeeds with the extreme complexity of natural scenes.

6.4. Sensory Processes

6.4.1. *Slow Chromatic Adaptation*

Adaptation to illuminant change includes processes with time constants on the order of minutes (Hunt, 1950). Significant changes of hue and saturation continue for at least 5 minutes. These shifts of unasserted color are in the general direction required to hold unasserted colors constant under illuminant change, but the quantitative accuracy of the compensation has been open to question. Two recent experiments (Arend, 1993a; Fairchild & Lennie, 1992) have shown definitively that the unasserted color shifts of slow chromatic adaptation fall far short of those required to compensate for even small illuminant changes.

In viewing multiple illuminant scenes we move our gaze from illuminant to illuminant at rates too high for slow chromatic adaptation to keep up (Judd, 1940). Although the momentary state of chromatic adaptation need not be uniform across the retina, the time constants for large adaptation changes are long enough that eye movements will frequently place scene regions under one illumination on portions of the retina adapted to another. In the Arend and Reeves (1986) chromatic Mondrian experiments, the subjects looked back and forth between continuously presented standard (6500 K illuminant) and test (4000 K illuminant) Mondrians, once per 2 seconds. Chromatic adaptation was too slow to equate the hues and saturations of identical-surface-color patches in our two simultaneously visible illuminants.

6.4.2 Rapid Chromatic Adaptation

Whereas chromatic adaptation cannot reliably produce illumination-invariant unasserted colors, it may still play a crucial role in surface-color perception. As noted earlier, rapid adaptation is required in achromatic vision to ensure that retinal responses encode local luminance contrast adequately. Neurons in the vertebrate retina need the rapid gain changes of retinal adaptation. Without them the neural responses would catastrophically saturate in reacting to the large luminance range in natural scenes (Shapley & Enroth-Cugell, 1984). Adaptation does not allow perfect encoding of achromatic contrast. At low luminances local apparent contrasts (brightness contrasts) depart significantly from Weber's law, and lightnesses are slightly affected (Arend, 1993b).

It is possible that chromatic adaptation also serves accurate encoding of image contrasts by chromatic mechanisms, and that departures from Weber's law affect chromatic appearance in a manner analogous to the achromatic case. However, recent experiments (Hayhoe, Wenderoth, Lynch, & Brainard, 1991) indicate that both the multiplicative and subtractive components of adaptation require several seconds to reach significant levels, much slower than the corresponding achromatic processes.

6.4.3 Simultaneous Contrast

Achromatic simultaneous contrast and gradient illusions also have their chromatic counterparts. When a gray patch is moved from a black background to a white one under the same illumination, both its brightness and its lightness change by about 1.5 Munsell Value steps. Unasserted colors depend on the chrominance of the immediate surround (Hasegawa, 1977; Kinney, 1962, 1965; Valberg, 1974; Ware & Cowan, 1983). In the classical textbook illustrations of chromatic contrast, the unasserted color of the disk surface is affected by the physical color of the surrounding surface. I know of no data as yet, but it is possible that apparent surface color is also affected by background color, producing failures of surface-color constancy.

Chromatic gradient illusions also produce significant distortions of unasserted color (Arend, 1973; Neri, 1989; Ware & Cowan, 1983). The illusions may also affect apparent surface color, producing departures from color constancy.

7. SUMMARY

Approximate contrast coding in the retina is not in itself sufficient to compute accurate apparent surface colors in natural scenes. The retinal image is an inherently ambiguous projection of the external physical world. To construct a reliable model of that world the visual system must supplement the information in the image with assumptions about the structure of the world. I suggest that the required assumptions are so elaborate as to constitute a kind of *psychological physics* with structure resembling, but not identical to, that of formal physics. Early visual processes are too simple to reliably invert the complicated physical interactions among lights, shapes, and materials.

The phenomenal appearance of the scene before our eyes suggests that the visual system computes iconic arrays of the surface properties on the way to its model of recognized 3D objects. Surface-color illusions indicate that the several perceived properties of surfaces are not computed independently. For example, in bistable patterns a change of perceived orientation can produce simultaneous changes of perceived surface color and illumination, changes interconnected in much the same way that the corresponding physical variables in the scene are related.

My proposed heuristic model of human surface-color perception poses questions about what quantities are explicitly represented in our perceptions, how they are computed, and how experimental data might be used to test surface perception algorithms.

According to this model the role of sensory mechanisms in surface-color constancy is to provide signals that approximately specify local luminance (and, possibly, chromatic) contrasts over a large dynamic range. Several experiments were discussed that addressed a question suggested by the model (i.e., how constancy of computed surface colors is affected by well-known departures of early visual mechanisms from ideal behavior).

ACKNOWLEDGMENT

This chapter was supported by the Air Force Office of Scientific Research (AFOSR 86–0128, AFOSR 89–0377).

5 Achromatic Transparency

Walter Gerbino
University of Trieste

1. INTRODUCTION

The relevant property of perceived transparency is the seeing of one color through another. This phenomenon contradicts the constancy hypothesis (i.e., the assumption that achromatic color corresponds to the local intensity). The obvious discrepancy between perceptual multidimensionality and stimulus unidimensionality accounts for the interest for transparency phenomena shown by many theorists.

Helmholtz (1866/1962) denied that the "seeing through" can be a matter of immediate sensory perception and claimed that an act of judgment is needed to separate two superposed colors. In his *Handbook* (Engl. ed. 1962, vol. II), Helmholtz considered transparency in the context of color mixture (pp. 120–122) and contrast (pp. 264–299). Color mixtures on partially reflecting surfaces, constancy of surface color, and shadows were explained within the same cognitive framework. The recognition of component colors would depend on a judgment "we are accustomed and trained to form" through experience with the same bodies seen under various illuminations (p. 287). According to this explanation, the observer acquires knowledge about color mixtures through the normal interaction with a world in which constant surfaces reflect variable light because the illuminant tends to change. Then, such acquired knowledge is used to disambiguate more complex mixtures, like those produced by colored mirrors and filters. More recently, Rock (1983) applied the problem-solving analogy to pictorial transparency. He showed how the transparency percept eliminates coincidental features of the stimulus and therefore represents a preferred solution.

Gestalt psychologists studied transparency as a kind of "duo formation" or "double representation" (Koffka, 1935). They rejected the idea that it is the result of judgment and studied it as a true percept, emphasizing its theoretical importance as a typical demonstration of structural organization in vision. According to the gestalt approach, stimulus conditions and autonomous processes of organization can account also for visual properties that are not in strict correspondence with local stimulus properties.

Gibson (1975, 1979) stressed the role of a proper description of transparent objects in the context of the ecological approach to visual perception. He suggested that transparent objects provide information about depth and therefore contribute to the solution of visual problems, more than posing a problem.

An adequate description at the level of ecological optics is essential to the understanding of vision. The next section is devoted to such a preliminary step.

1.1. Optical Ecology

Our environment contains several kinds of transparent things, and we perceive transparency in a variety of conditions. First, consider the following classification into media, substances, and layers, before focusing on layers, which have been modeled and studied psychophysically.

Media. Perception of ambient illumination represents a ubiquitous case of transparency. The whole history of the concept of "light" (Ronchi, 1983) suggested that the emptiness between surfaces and bodies is perceived as evenly filled by a diaphanous medium, having at least one property, intensity. However, it has been questioned whether illumination as such is immediately perceived, in a clear and homogeneously lighted environment.

We certainly perceive denser media like air at a distance, fog, or water (when diving), as well as the nesting of different illuminations in a rich environment, containing multiple sources and shadowed zones.

Transparent media appear to fill the space between bodies and act as their three-dimensional background. This property differentiates them from substances and layers.

Substances. Things like clouds, liquids in containers, and blocks of glass often are and appear to be transparent. Some exhibit a very mutable shape and can even interpenetrate each other; some do not. But this is rather irrelevant to the basic experience of "seeing through."

A transparent substance is a visual entity that occupies a specific volume, has its own color, and largely preserves the structure of background objects. If alterations of the geometrical structure of background contours occur, they can be veridically attributed to the distorting power of the transparent substance. This happens in some cases of physical refraction. In other cases, the disruption of background contours by refraction is not attributed to the transparent surface.

Layers. For theoretical reasons, it is convenient to consider together several kinds of transparent strata. Gibson (1975) maintained a distinction among sheets, meshes, and shadows. They differ in several respects, but a single physical model, the layer, can account for all.

Ordinarily, the alternation of holes and solid material is not visible in sheets (i.e., films, panes, mirrors, or filters), whereas it is visible in meshes (i.e., grids, nets, or veils). The former look homogeneous; the latter look textured. But, if one disregards differences due to matte versus mirror reflection on the front surface, the distinction between sheets and meshes merely becomes a function of observation distance; that is, when fibers cannot be resolved, meshes become indistinguishable from sheets. The same is true of episcotisters (i.e., wheels, fans, or propellers), which modulate in time what meshes modulate in space. A blade rotating at fusion speed is equivalent to, and looks like, a veil.

Shadows differ from sheets and meshes because they always appear attached to a background surface, whereas sheets and meshes can be seen clearly detached. However, it is important to notice that shadows look like thin immaterial layers.

These three kinds of layers are physically different but produce similar optic arrays. Their common optical properties appear clear when we consider the achromatic domain and try to describe the corresponding mixtures of light intensities. Consequently, a single theoretical framework, one that addresses intensity relationships, in principle can allow us to understand perception of both material and immaterial layers.

2. LAYER TRANSPARENCY AND COLOR SCISSION

A layer is a convenient object for the study of perceived transparency because its underlying physical model is particularly simple. Strange as it might seem, two different models are present in the literature. The first is the *episcotister* model, proposed by Metelli (1970) and extensively studied by him and his associates. The second is the *filter* model, described by Beck, Prazdny, and Ivry (1984) and Brill (1984). Both models do share the same fundamental assumptions about the idealized layer. They differ, however, with respect to implied boundary conditions.

2.1. Layer Physical Model

Preliminarily, let us clarify two assumptions about the idealized layer that make it useful in the study of both shadows and transparent layers:

Thickness. First, a layer is conceived as an arbitrarily thin wafer of two surfaces (imagine the upper and lower sides of a sheet of paper). In a matte achromatic world, each surface is completely described by a combination of three values that quantify the proportion of incident light that is transmitted, reflected, or absorbed. Because the three proportions sum to unity, we need only specify transmittance and reflectance, leaving absorbance as a residual.

Symmetry. Second, a layer is generally conceived as symmetrical (i.e., both surfaces of the wafer are described by the same pair of values). Let us anticipate that this assumption is irrelevant for the episcotister model but relevant for the filter model.

In a world of achromatic, homogeneous, textureless, and matte surfaces, the optic array generated by superposing a layer upon opaque background surfaces is represented by a set of intensity values. Figure 5.1 illustrates this prototypical situation and demonstrates that transparency can be pictorially displayed by a well-chosen set of intensities, within the range of paper reflectances. It should be noted that figural conditions, as well as observer's attitude, play an important role in pictorial transparency. In impoverished displays, however, one should always distinguish between two sets of conditions: those determining the occurrence and maintenance of a given organization, and those controlling the internal properties of organized objects, when they are phenomenally instantiated. It is controversial whether intensity conditions belong to the first set (Beck et al., 1984; Metelli, Da Pos, & Cavedon, 1985). However, intensity conditions certainly belong to the second, as shown by experimental evidence indicating that the degree of perceived transparency is lawfully related to intensity relationships, under constant figural conditions (Metelli et al., 1985).

Episcotister and filter models are directly concerned with color conditions. Both refer to the four-intensity pattern illustrated in Fig. 5.1 and obey (although not always explicitly) the following logic. First, one describes the optic array resulting from the superposition of a layer on a background made of two adjacent opaque surfaces, under given illumination constraints. This step has to do with distal-to-proximal mapping (i.e., with the

5. ACHROMATIC TRANSPARENCY   219

FIG. 5.1.   Pictorial transparency in a four-region pattern.

generation of the pattern of four luminances *ABPQ*; Fig. 5.2). It consists in the physical description of the optical transformation of background luminances as a consequence of light filtering by the layer. Second, the inverse problem is formulated: Can distal properties of the layer be recovered by the visual system on the basis of proximal information, and how? The proposed answer is: Yes, and this is done precisely by reversing the distal-to-proximal mapping and using it as a model of the proximal-to-perceptual mapping. The visual system would have the capability of splitting *P* and *Q* into layer and background components.

The term *scission* has been selected to describe the partitioning of a local intensity into several components (Kanizsa, 1979; Metelli, 1975). This term usefully describes the ubiquitous partitioning of a local luminance into illumination, on the one hand, and surface color, on the other; a condition where, according to gestalt formulations, the illumination component is the *framework* for the surface component, which is the *thing* (Koffka, 1935). The same term describes the partitioning into two surface components occurring in transparency, the one between the front layer surface and the background surface.

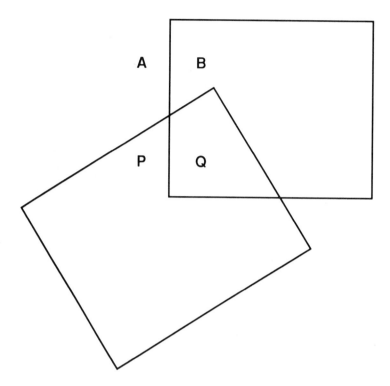

FIG. 5.2. The four-luminance pattern *ABPQ*. The layer transforms *A* and *B* luminances into *P* and *Q* luminances.

Conceptually, scission is the opposite of fusion. Fusion refers to distal-to-proximal mapping and describes all kinds of color mixtures. One can say, for instance, that incident illumination and surface reflectance "fuse" into a single local luminance.

Fusion and scission are used here only in a metaphorical sense. One must be aware that scission, in particular, can suggest that a primary sensation is split into components by a secondary perceptual process, which is not the intended meaning. In this chapter, it is a descriptive term for a particular organizational outcome, without any reference to an actual process. Abstractly, scission corresponds to vector decomposition (Johansson, 1982) or movement apportionment (Rock, 1983) in the organization of motion trajectories. This correspondence between color and motion domains has been analyzed by Bergström (1982).

According to episcotister and filter models, a proximal-to-percept mapping (perceptual scission) is the reverse of a distal-to-proximal mapping (optical fusion). This theoretical proposal is flawed. It suggests that distal properties can be reliably recovered from proximal information. Yet optical information is notoriously burdened by local undeterminacy. Each of the four luminances results from several components; in particular, luminances corresponding to the layer are the mixture of three object components (background reflectance, layer reflectance, and layer transmittance), plus an illumination component. Locally, the distal-to-proximal mapping is many-to-one; hence, reversing its direction does not seem very promising. For instance, one cannot determine surface color only on the basis of the corresponding luminance, unless the illumination component is defined. The discussion of episcotister and filter models in the luminance domain will hopefully clarify this criticism further.

Before going into that matter, however, one general point should be emphasized. Perceived transparency in the four-intensity pattern is especially interesting, because its occurrence is consistent with the hypothesis that color scission depends on an autochtonous tendency to simplicity (i.e., that the hierarchical organization of illumination, transparent layers, and opaque background is constrained by the minimum principle; Hatfield & Epstein, 1985; Kanizsa & Luccio, 1986). The next section deals with this aspect.

2.2. Hierarchical Organization

In this section, I illustrate how the minimum principle can be used to interpret organization occurring in transparency. I refer to Structural Information Theory because it provides a general framework for comparing the complexities of different solutions (Leeuwenberg & Buffart, 1983).[1]

Consider first several structural descriptions of the four-luminance pattern, and the abstract notion of scission. In this context, scission is conceived as a mapping such that any luminance (A, P, etc.) corresponds to one illumination component (I) and some surface components: Any surface is described by two parameters, reflectance (a, p, etc.) and transmittance (t). An opaque surface is indicated by null transmittance. Given that, one can conceive of innumerable partitionings of a four-luminance pattern into common and individual components. Figures 5.3 to 5.6 illustrate significant ones. The complexity of each partitioning is obtained by counting the number of parameters involved in the description and constitutes

---

[1]Leeuwenberg may disagree on some details of this particular application of the coding approach. However, it follows general principles of hierarchical coding already applied to other domains, as motion trajectories (Restle, 1979) and pattern contours (van der Helm & Leeuwenberg, 1988). Previous applications of the coding approach to transparency and to the "neon illusion" (Leeuwenberg, 1978; Tuijl & Leeuwenberg, 1979) were concerned with figural aspects, not with intensity relationships.

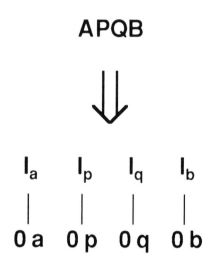

FIG. 5.3. Local partitioning of the four luminances.

the information load of the organization. The symbol C (complexity) is used to denote information load, just to avoid confusion with the symbol I, used here for the illumination component.

Partitioning 1. Each of the four luminances is locally partitioned and coded independently from the others. This gives rise to the most complex solution, which would occur if scission were not constrained by the minimum principle. One never perceives according to this partitioning, which theoretically implies that each surface is seen as having a different color and its own illumination, different from those falling on adjacent surfaces. Because each local partitioning involves three parameters, the information load of four independent partitionings is C=12. Of course, one could think of innumerable superpositions of transparent surfaces, leading to even more complex solutions with an infinite number of entities and parameters. To stop this recursion, let us impose a default value of 0 to transmittance.

Partitioning 2. An illumination component I, common to all luminances, is extracted. Displays containing two or more luminances always allow the extraction of an arbitrary common component. Classical experiments on two-luminance patterns (Wallach, 1948, 1976) indicate that lightness perception occurs in accordance with this simple kind of organization.

Partitioning 2

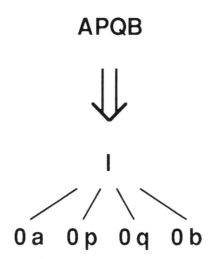

FIG. 5.4. Partitioning of the four-luminance pattern into a single common illumination and four independent opaque surfaces.

In the four-luminance pattern, such a partitioning corresponds to perceiving a mosaic of four differently colored surfaces under a common illumination. Formally, the extraction of a common illumination component is consistent with Wallach's ratio principle of adjacent luminances. When a common illumination is factored out, ratios between residual reflectances are equal to luminance ratios. Because luminance ratios are invariant under varying common illumination, they can provide a basis for color constancy in the normal environment. With more than two luminances, the extraction of a component that is common to all of them corresponds to *edge integration* (Gilchrist et al., 1983). This solution has an information load $C = 9$ that, relative to the previous partitionings, is a clear improvement in terms of coding economy.[2]

Partitioning 3. Four proximal regions are mapped into three surfaces. As it happens when looking at Fig. 5.1, one transparent surface is seen on top

---

[2]One may wonder why the 0, that describes opacity of each region, is counted four times, instead of being grouped and counted only once. This would be the consequence of a mechanical application of the alternation rule to the sequential pattern *I0aI0pI0qI0b*, which could be coded as <*I0*><*apqb*>. A semantic constraint should prevent ungrouping of parameters that define surface properties. Remember the basic assumption that a surface needs to be defined by two parameters.

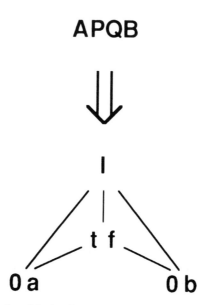

FIG. 5.5. Organization of the four luminances into a transparent layer superposed on a two-color background, under homogeneous illumination.

of two adjacent background surfaces. This partitioning is constrained by luminance relationships. Depending on the actual intensities, the four-luminance pattern can or cannot fit into this hierarchical structure. If it fits, the simplest hierarchical organization is achieved, in which a common illumination is distributed over a transparent layer and two opaque surfaces. The savings of Partitioning 3 upon Partitioning 2 requires at least four well-chosen luminances. No simplification is possible if the pattern contains only three luminances, whatever they are. Hierarchical organization of the *APB* set according to Partitioning 2 is *I* [(0*a*)(0*p*)(0*b*)], with information load $C = 7$. Complexity is not reduced if scission occurs according to Partitioning 3, and one region is transparent; consider, for instance, the mapping of the *APB* set into *I* [(0*a*)(*tf*)(0*b*)]. This is not to say that transparency cannot be perceived in a pictorial display containing only three reflectances. This outcome may occur (Masin, 1984) but cannot be explained by an autochtonous tendency to the simplest organization of color; the application of this principle to intensity relationships, independently from figural conditions, requires at least four distinct values. Partitioning 3 is the most economical partitioning of a four–luminance pattern according to

Partitioning 4

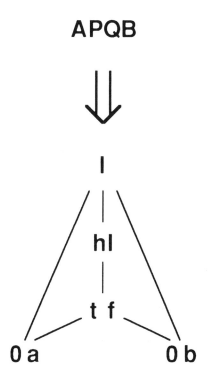

FIG. 5.6. Organization of the four luminances into a transparent layer superposed on a two-color background. The layer illumination is different from the background illumination.

previous assumptions; the corresponding code has an information load $C = 7$.

Partitioning 4. In Partitioning 3, the code represents illumination as homogeneously distributed over both layer and background surfaces. In ecological optics, this is a limiting case. Partitioning 4 represents a more general case, in which layer illumination hI is not necessarily equal to background illumination I.[3] A nesting of the two illuminations is embodied

---

[3]Think of a thin curtain in front of a window, in a normally illuminated room. At noon, the outdoor scene seen through the veil by an inside observer is more illuminated than the veil; in the evening, it is less illuminated. Under these circumstances, the assumption of homogeneous illumination works just like a broken watch: It happens to be correct twice a day.

in the structure because of its ecological representativeness; namely, the more local layer illumination is coded as a multiple of the including background illumination I. However, considering the two illuminations as either independent or nested does not affect complexity. This solution has an information load $C = 8$ and, despite the presence of two illuminations, is simpler than the less articulated Partitioning 2. Partitioning 4 is compatible with both the episcotister and the filter models, although their proponents never considered it. Because of its generality, this partitioning has to be kept in mind when evaluating experimental evidence contrary to restrictive formulations of algebraic models of transparency.

In sum, different partitionings have different complexity. A pattern made of at least four luminances allow two partitionings involving a layer (3 and 4) that are more economical than partitionings not involving a layer (1 and 2). In the four-luminance pattern, the unification of two regions into a single layer and the partitioning of corresponding luminances into layer and background components are supported by the minimum principle.

2.3. Levels of Scission Theory

Let us distinguish three levels in the scission theory of transparency: structural, metrical, and physical. The first level is more general than the second, which in turn is more general than the third:

Structural level. The structural level regards the hypothesis that scission of a four-luminance pattern occurs according to the minimum principle. The most economical partitioning is preferred because of the tendency toward the simplest organization. If intensity relationships allow it, perception will occur according to Partitioning 3. The hypothesis that scission follows the minimum principle is very general and does not specify the partitioning function. Therefore, it does not explain how to find the exact values of layer parameters, given the four luminances. In principle, infinite solutions of Partitioning 3 are possible.

Metrical level. The metrical level is connected with the choice of a specific algebraic model, which permits quantitative predictions. Given a set of local luminances, layer properties are derived according to scission functions.

Physical level. The physical level regards the environmental state of affairs that one may have in mind to choose among alternative algebraic models. It involves a spinning episcotister, a filter, or any other material layer, together with some explicit and implicit assumptions about the reciprocal positions of layer, background, illumination source, and viewpoint. These assumptions are the boundary conditions of the physical model.

Logically, the choice of a given physical model implies a given algebraic model, which in turn implements some kind of structural organization. However, each more general level is independent from less general ones. An algebraic model could be preferred quite apart from its physical plausibility. And the notion of structural organization in transparency could still be valid, even though at a given time one might be unable to define its exact algebraic form.

As a matter of fact, current algebraic models of scission in the four-intensity pattern are tied to specific physical models. Hence, my discussion of episcotister and filter models, also referred to as Metelli's and Beck's models, considers both aspects together.

According to their proponents, both models allow computation of layer reflectance and transmittance using only the four intensities as independent variables, and predictions of perceptual scission and appearance of the layer (i.e., its color and degree of transparency). As I show later, however, the same predictions require fewer assumptions in the episcotister model than in the filter model.

Both models express light intensities as reflectances, a theoretical choice that poses a quandary for those who are convinced that the eye is not in touch with distal properties. Recently, both Beck and Metelli have considered the possibility of using subjective measures (i.e., lightnesses) instead of reflectances, to account for inadequacies of original formulations (Beck et al., 1984; Metelli, 1985). However, I feel that one goes too far from the observer when using reflectances but comes too near when using lightnesses. If the basic concern is optic information, on the one hand, and perception, on the other, then the right choice is luminances. I review each received model following its original reflectance-based formulation and then introduce the luminance-based modification.

A last warning is necessary. Unfortunately, neither Beck et al. (1984) nor Brill (1984) used the same notation used by Metelli and his co-workers through a long series of papers. For the sake of comparison, I here introduce a uniform notation (as adherent as possible to Metelli's original one) that is used in describing both the episcotister and the filter models, as well as in developing new analyses.

Uppercase letters represent light intensities; for instance, luminances A, P, or incident illumination I. Lowercase letters represent rational numbers; for instance, reflectance values a, p, r, or transmittance t. Because no ambiguity ordinarily arises when describing spatial arrangement, regions are denoted by lowercase letters as well. The four intensities refer to a spatial arrangement of regions containing adjacencies *ap*, *pq*, *qb*, and *ab*. Conventionally, scission functions are applied under the assumption that the *ab* pair corresponds to the background and the *pq* pair to the layer. However, the mapping of the four regions into one layer on two background surfaces is not univocal. Three kinds of four-intensity patterns exist: those with no partitioning solution; the unambiguous ones, with just one solution; and the ambiguous ones, with two solutions.

3. THE EPISCOTISTER REFLECTANCE MODEL

The cornerstone of Metelli's model is the idea that scission is the reverse of fusion obtained by rotating an episcotister (Heider, 1933; Koffka, 1935). An episcotister is a rotating disk that alternates open and solid sectors. Helmholtz (1866/1962, II, p. 199; III, p. 522) referred to Talbot as the first to use it and described its convenience for the accurate fractioning of the intensity of light sources.

Metelli referred to color fusion in a physical situation in which an episcotister rotates in front of an opaque background of reflectance *a*; the episcotister has an open sector of size *t* (a proportion of the total disk), and a solid sector of size $1 - t$ having reflectance *r*. Then, *p* is used to indicate what he calls the "reflectance of the resulting mixture" (Metelli, 1970, p. 61) or "reflectance of the fusion color" (Metelli et al., 1985, p. 354); a concept with a rather uncertain status. One may prefer to say that *p* is the reflectance of a solid surface that perfectly matches the rotating episcotister. Following Talbot's law, Metelli proposed the following equation for *p*:

$$p = ta + (1-t)r \qquad (1)$$

Virtual reflectance *p* is equal to the weighted sum of background reflectance and episcotister solid sector reflectance. It is never made explicit that the validity of Equation 1 depends on the distribution of incident illumination. If and only if the illumination falling on these surfaces is perfectly homogeneous (i.e., if the same amount of light falls on both the occluded background and the episcotister), then fusion by an episcotister is equivalent to fusion by a fully solid disk divided into sectors of different reflectances (a Maxwell disk), for which Equation 1 holds without further restrictions. The reason for disregarding illumination could be strictly dependent on the use of pictorial displays viewed under homogeneous illumination. However, one should never confound the illumination under which a picture is viewed with the illumination or illuminations represented in the depicted scene.

One may raise the question whether the law expressed by Equation 1 is observer-dependent or observer-independent. Helmholtz (1866/1962, II, section 22) devoted a whole paragraph to the problem of intensity mixtures obtained by intermittent stimulation and treated it as an empirical problem of physiological optics; hence, considering Talbot's law as observer-dependent. He evaluated experimental evidence about spinning disks with sectors of different reflectances (not episcotisters) and came to the following general conclusion: "When a certain place on the retina is stimulated always in the same way by regular periodic impulses of light, then, provided each recurrent stimulus is sufficiently short-lived, the result is a continuous impression, equivalent to what would be produced if the light acting during each period were uniformly distributed over the entire time" (Helmholtz, 1866/1962, p. 207). Because of the association with Talbot and the photographic process, one can better claim that Equation 1 expresses a law of physical optics, at a gross temporal scale; which is consistent with the notion that, in visual theory, fusion and scission are metaphors, not actual processes.

When the episcotister is superposed on two background reflectances a and b, two virtual reflectances result: p and q. The following system of two equations applies to $apqb$ reflectances:

$$p = ta + (1-t)r \qquad (2)$$

$$q = tb + (1-t)r \qquad (3)$$

The solutions for t and r are:

$$t = \frac{p-q}{a-b} \qquad (4)$$

$$r = \frac{aq - bp}{(a+q) - (b+p)} \qquad (5)$$

In the reflectance version of the episcotister model, all variables represent rational numbers belonging to the [0,1] domain: t is the proportion of open sector, and a, p, q, b, and r are reflectances, that is proportions of reflected light. Given these specifications and constraints on adjacent regions mentioned earlier, straightforward predictions about the occurrence of scission and the appearance of the transparent layer are possible:

Occurrence. Given four reflectances $abpq$, homogeneous scission of pq will occur and support the perception of balanced transparency if and only if both t and r belong to [0,1]. When this condition is not fulfilled, one of two alternatives should take place: Either no scission occurs and the display is perceived as a set of four independent colors (Partitioning 2 in Fig. 5.4), or pq is inhomogeneously partitioned, giving rise to some kind of unbalanced transparency, in which each layer region has its own appearance. The received episcotister model is meaningless outside the intersection of $[0 \leq t \leq 1]$ and $[0 \leq r \leq 1]$.[4]

Appearance. Within the previously defined domain, perceived transparency should be a function of t and r. Metelli proposed an interpretation very close to the physical model underlying Equations 2 and 3. He hypothesized that perceived transparency is two-dimensional. Thus, there should be two phenomenal dimensions of the layer: apparent transmittance, which is a function of t, and color, which is a function of r. But Metelli also considered the possibility that the perceived degree of transparency is a unitary psychological dimension, which is a joint function of both parameters.[5]

3.1. Episcotister Constraints to the Scission Function

Regardless of the problem of dimensionality of perceived transparency, one can focus on the constraints connected with the meaning of parameters of the scission function. To facilitate later comparison with the filter model, let us modify Metelli's formulation. Let f, where $f = (1 - t)r$, be a common additive component and rewrite Equations 2 and 3. The resulting model and the f solution are as follows:

$$p = ta + f \qquad (6)$$

$$q = tb + f \qquad (7)$$

$$f = \frac{aq - bp}{a - b} \qquad (8)$$

Consider the elements of Metelli's theory and their logical relationships. One can examine Equations 5 and 6 from two different perspectives: either

---

[4]Metelli et al. (1985) stated: "If transparency occurs outside the afore-said limits, then the transparency is unbalanced, and the theory makes no assertions about the possible existence of forms of nonbalanced transparency" (p. 355). This conclusion is valid if and only if the assumption about homogeneous illumination is made explicit; otherwise, as I suggest later, any positive value of r is allowed. In general, one should always distinguish between the model, and what can be derived from it, and the phenomenal occurrences of transparency. By ad hoc assumptions, equations that look similar to Equations 4 and 5 have been suggested to account for unbalanced transparency (Metelli, 1975), but they cannot be derived from any physical model.

[5]Metelli (1976) reported paradoxical judgments, especially by naive observers, in a task where one has to choose the more transparent display out of two. Referring to the two-dimension hypothesis, he explained inconsistent results by shifts of attention from one dimension to the other. Which hypothesis is better remains an open empirical question.

5. ACHROMATIC TRANSPARENCY

as a scission function or as a symbolic mapping specific to the episcotister arrangement. The scission perspective emphasizes the abstract problem of magnitude partitioning. The symbolic mapping perspective regards the way a particular physical arrangement maps onto a proximal configuration.

Take the first view, which is the more general one. Equations 6 and 7 correspond to a simple version of the partitioning problem, which may be stated as follows: "Given four magnitudes $apqb$, subtract from both p and q an identical magnitude f (if it exists) so that residual magnitudes are proportional to a and b." According to this idea, the common component f cannot be larger than the smaller of the two values p and q. The latter constraint is due to the fact that the individual components ta and tb, as residuals in subtraction, cannot be negative because these correspond to physical magnitudes.

Take now the second view, which is Metelli's physical model (i.e., a particular spatial arrangement of episcotister, background, and illumination). To understand the meaning of the additive component, let us introduce the distinction between effective reflectance and material reflectance. In the preceding expressions, the common component f refers to effective reflectance, and r (the weighted factor of f) refers to material reflectance. Effective reflectance f is the proportion of light reflected by the layer as a whole. Material reflectance r is the proportion of light reflected by its opaque portion, covering only a fraction of the total area; it is the reflectance of solid blades of rotating episcotisters, or the reflectance of opaque particles of static layers (like veils or filters). This distinction is an important one, but it appears to have been overlooked by Beck et al. (1984, p. 415) in their comparison of episcotister and filter models. The underlying physical model implies that f cannot be larger than $(1-t)$; which is another way of saying that r cannot be larger than 1.

In sum, if we examine Metelli's theory from two different perspectives, we arrive at respectively different conclusions. According to the scission idea, f cannot be larger than p; however, according to the episcotister model, f cannot be larger than $(1-t)$. The second constraint is more restrictive than the first. This fact has important implications to be considered in more detail in the next section. To preview briefly, the $f = (1-t)$ or $r = 1$ boundary can be shown to exclude certain sets of four-reflectance displays as reliable situations for transparency percepts simply because they do not match the underlying physical model, and not because they fail to allow partitioning into common and individual components. Such a criticism is consistent with the depiction of levels of scission theory discussed in section 2.3. In the next section this and related points are analyzed visually using graphical representations of reflectance values.

3.2. Graphical Representation

Let us pose the following question: Given a pair of background reflectances *ab*, how should two other reflectances *pq* be selected so that *p* and *q* are partitionable, and therefore the resulting *apqb* pattern generates a percept of transparency? This question is essentially at the heart of the transparency issue in the pictorial domain. From the point of view of a painter, for instance, it is important to know how to choose the two grays that, in combination with a given two-color background, will be perceived as a homogeneous veil.

In the attempt to address this problem, Remondino (1975) provided a convenient graphical representation of the episcotister model. In general, his method permits visualization of relations between four terms, two constants (here *a* and *b*) and two variables (here *p* and *q*); two terms are kept constant to allow the construction of an easy two-dimensional diagram. Therefore, one has to draw a different diagram for each background pair. Figure 5.7 shows the two-dimensional reflectance space, which is the unit square [(0,0)(0,1)(1,0)(1,1)], with an arbitrary (*b,a*) point chosen to represent an *ab* pair of background reflectances. Conventionally, the *a* value is located on the vertical axis and the *b* value on the horizontal axis.

According to Metelli's model, the set of (*q,p*) points corresponding to valid solutions will be contained within a given region of the reflectance space. Notice that *p* is chosen on the same axis as *a* and *q* on the same axis as *b*. The solution region is bounded by four straight lines, each of which corresponds to one limit of the domains for *t* and *r*: $t = 0, t = 1, r = 0, r = 1$. This region is actually a triangle (shaded in Fig. 5.7) because three boundaries ($t = 1, r = 0, r = 1$) intersect each other in the (*b,a*) point.

Consider now the four boundary lines. Let us rely on the general abstraction of $y = f(x)$ to express the way in which $p = f(q)$ from Equations 2 and 3. Solving these equations for each of the four limits we obtain four straight lines:

$t = 0$: $y = x$; this is the main diagonal from (0,0) to (1,1) corresponding to all cases where $p = q$;

$t = 1$: $y = x + (a - b)$; this specifies a parallel to $y = x$ through the (*b,a*) point, where $p = q + (a - b)$;

$r = 0$: $y = (a/b)x$; this line passes through the origin and the (*b,a*) point; it specifies the cases where $p/q = a/b$;

$r = 1$: $y = ((1-a)/(1-b))x + (a-b)/(1-b)$; this line passes through (*b, a*) and (1, 1) points.

5. ACHROMATIC TRANSPARENCY 233

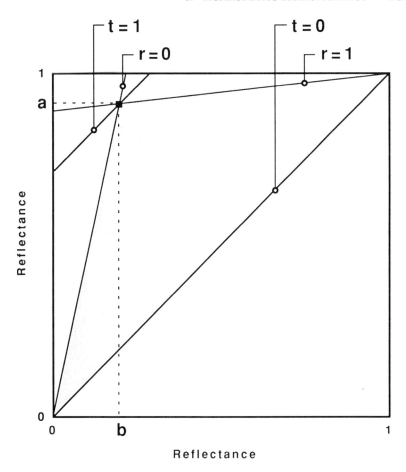

FIG. 5.7. Diagram in the reflectance space in which the shaded area represents the valid set of *pq* pairs, relative to the specific *ab* background pair. The shaded area is bounded by four straight lines, corresponding to *t* and *r* constraints.

The shape of the triangular region delimited by $t = 0$, $r = 0$, and $r = 1$ is contingent on the choice of the *ab* pair. However, some geometrical features make this representation a useful tool for graphically finding $t, f,$ and r solutions graphically, given an arbitrary (q,p) point.

The graphical representation permits identification of t as a ratio, a dimensionless number. This is shown in Fig. 5.8. Note that here the distance between the two boundaries $t = 0$ and $t = 1$ defines a linear scale of unit length. Thus, the distance from a specific point to the $(t = 0)$ line defines the related t value as a proportion. This can be shown by observing

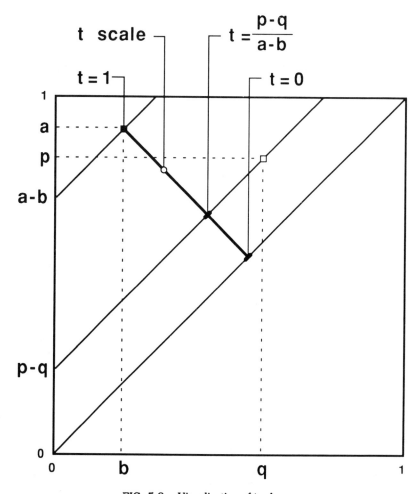

FIG. 5.8. Visualization of t value.

that the t value is the ratio between two (vertical) intercepts, $p-q$ and $a-b$, associated with lines respectively through (q,p) and (b,a), along which t is constant.

The f value is represented graphically as shown in Fig. 5.9. The key idea shown here is that f appears as a common component to be subtracted from both p and q. The f value thus is shown as a segment along each axis (vertical and horizontal). To see this, notice that the point where f is zero for a given pq pair is defined by the intersection of the t line through the (q,p) point and the boundary line $r = 0$. This intersection point (S in Fig. 5.9) projects a null f value to both vertical and horizontal axes; f segments in p

5. ACHROMATIC TRANSPARENCY   235

and q originate from these points and extend to p and q, respectively. Hence, the simultaneous partitionings of both p and q become visible on the two axes: Along the vertical, p is partitioned into $ta + f$; along the horizontal, q is partitioned into $tb + f$. This visualizes the idea that f is the magnitude one has to subtract from both p and q, to eliminate the difference between p/q and a/b, a difference that constitutes an imbalance between layer reflectance ratio and background reflectance ratio. Note that when the additive component f is nullified this imbalance disappears because $p = ta$ and $q = tb$; which means that the intensity ratios of layer regions and background regions are the same. Rearranging, this yields $p/q = a/b$, meaning that layer and background ratios are the same.

The graphical portrayal of r is especially interesting. Physically, r repre-

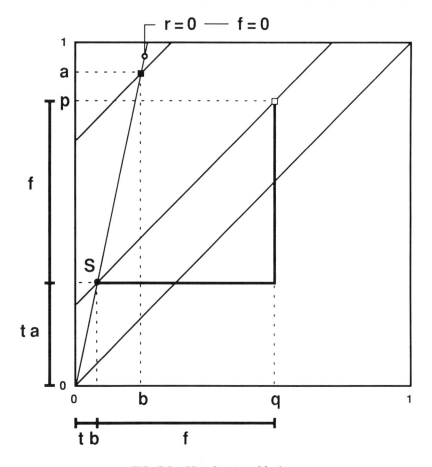

FIG. 5.9. Visualization of f value.

sents layer material reflectance and therefore it cannot assume values greater than 1. Graphically, violations of this constraint, if they occur, can be clearly depicted. To see this consider Fig. 5.10. Here r appears as a vertical (or horizontal) segment. Its coordinate value on a given axis is determined by a point on the boundary line $t = 0$. This point is the intersection of this boundary line with the straight line through (b,a) and (q,p) points. The graphical demonstration, based on triangle similarity, exploits equation $f = (1 - t)r$. Remember that in the diagram f and r are scaled relative to the unit of the reflectance space, whereas t is scaled relative to the distance between the (b, a) point and the $t = 0$ boundary.

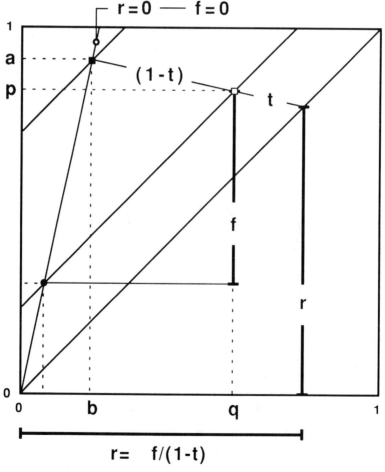

FIG. 5.10. Visualization of r value and graphical demonstration that $f = (1-t)r$.

5. ACHROMATIC TRANSPARENCY   237

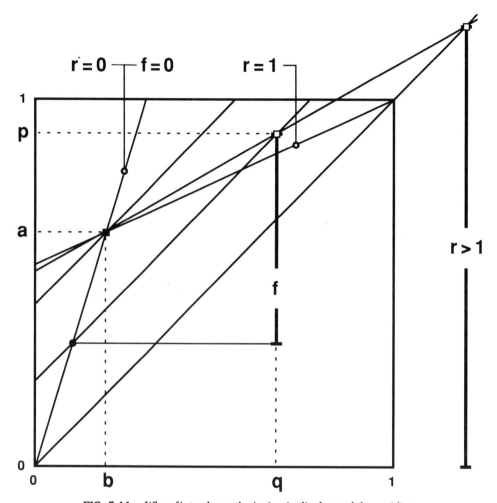

FIG. 5.11. When f is too large, the (q,p) point lies beyond the $r = 1$ line.

Figure 5.11 shows that partitioning of p and q into common and individual components is possible also when f is such that the (q,p) point lies beyond the $r=1$ line. As far as scission is concerned, this boundary is meaningless. The solution space is delimited by just three boundaries, intrinsic to the very idea of partitioning into common and individual components: $t = 0$, $t = 1$, and $f = 0$. Any other constraint is extrinsic to this algebraic formulation of scission. When the (q,p) point lies beyond the $r = 1$ line, the straight line connecting (q,p) and (b,a) intersects the main diagonal outside the reflectance space. This visualizes the fact that the physical element of Metelli's theory (the episcotister reflectance model) is more restrictive than the structural element (the scission hypothesis).

In addition to the theoretical criticism concerning the validity of the $r=1$ boundary, there is empirical evidence that transparency is perceived when the value of r, computed according to Equation 5, is larger than unity. In their extensive exploration of the transparency space, Beck et al. (1984, Experiment 4) found that transparency judgments are frequent even when $r>1$. For instance, a four-reflectance pattern corresponding to $t=.96$, and $r=3.38$ was accepted as transparent by 17 out of 26 observers; and another corresponding to $t=.88$ and $r=2.90$ by 20 out of 26. Consider that the same proportion was obtained using a perfectly admissible pattern with $t=.88$ and $r=.08$ (Tables 6 and 7, pp. 419–420). My observations with similar patterns (Gerbino, 1985) led to the same conclusion: The $r=1$ boundary can be violated, in a nontrivial way, without impairing perceptual transparency. But according to the episcotister reflectance model, these displays with $r>1$ should not be judged as transparent. Of course, in such patterns the perceived layer has a special appearance, and this can be easily predicted by having another look at Fig. 5.7. Generally speaking, when t has an admissible value and r is larger than unity, the layer is much brighter than the background. Under these conditions, the *kind* of transparency is clearly different from the one obtained when the layer is, on the average, darker than the background.

Metelli (1975a) himself noticed that transparency can be perceived even when the layer appears too bright: "There are also situations where the transparent layer is described as wet or polished—and therefore endowed with a sort of extra-reflectance" (p. 478).

All these facts point in the same direction. In its received form, the episcotister model cannot explain perceived transparency when the (q,p) point is outside the triangular shaded region of Fig. 5.7. Nevertheless, this percept does occur also with these illegal four-reflectance patterns. I claim that counterfactual evidence that has been reported should be expected, not because scission is an inadequate metaphor or because partitioning equations are incorrect but because Metelli's formulation is unnecessarily restrictive. Any formulation using reflectances as independent variables disregards illumination. The discussion of illumination constraints introduces us to the luminance episcotister model.

3.3. Illumination Constraints

Basically, there are three cases of uneven illumination that are incompatible with the received episcotister model. These cases apply not only to episcotister-generated veils but to any kind of layer; therefore, the general term *layer* is used here. Cases 1 and 2 contemplate conditions even more specific than those embodied in Metelli's model, whereas Case 3 is the generic one.

Case 1. Take the following arrangement. Objects and illumination source are located in such a way that the layer casts its own shadow upon the background, and layer and shadow coincide from observer's viewpoint. This can occur in two circumstances. First, when a directional source, like a projector, is located near the observer, and there is no other environmental illumination; the layer-to-background distance is irrelevant. Second, when the layer is virtually in contact with the background; here, the distinction between directional and diffuse illumination is irrelevant. Metelli (1983, p. 164) concerned himself with episcotister-to-background distance when considering depth information for the presence of a layer; a variable one needs to nullify to isolate color conditions. He suggested that distance can be minimized in such a way that the episcotister setting becomes virtually indistinguishable from a flat arrangement of paper reflectances. However, when distance is reduced and the cast shadow becomes significant, Equations 2 and 3 no longer describe the relevant optic array, because the implicit assumption of homogeneous illumination is violated. As a curiosity, perform the following *Gedanken* experiment. In a normally lighted room, judge the transparency of an episcotister rotating very near the background. If judgments agree with Metelli's predictions, we should experience a conflict. Data confirm the algebraic model, but they depart from veridicality (they do not match the reflectance and opening of episcotister blades); nevertheless, the model is tailored on the very device used in the experiment. When the episcotister is close to the background, Metelli's algebraic model becomes inadequate. In fact, the fractioning factor expressing the proportion by which light reflected by the background is transmitted through the layer cannot simply be equivalent to layer transmittance (t in Equation 2), but at best it is squared layer transmittance (t^2). This is fully clarified in the description of the filter model. When proposing the filter model, Beck et al. (1984) referred exactly, although implicitly, to such a physical state of affairs. Case 1 arrangements, of course, are not typical of filters but of any layer under these circumstances.

Case 2. The opposite extreme case of uneven illumination occurs when no light falls on the layer. All light reaching the eye comes from the background, either directly or through the layer, fractioned by a factor t. When no light is reflected by the layer, the four–intensity pattern corresponding to a layer is equivalent to that corresponding to a shadow, independently from layer reflectance. Metelli (1975b) analyzed shadows in connection with his algebraic model of transparency and noticed that intensity relationships arising from the superposition of an illumination edge over a reflectance edge are indistinguishable from those generated by an episcotister with a solid sector of virtually no reflectance rotating in front of the same reflectance edge (under homogeneous illumination, one should add). The same happens when the episcotister, even a white one, is not illuminated. The multivocity of the limiting case $[t = p/a = q/b]$ becomes more apparent when the whole model is reformulated using luminances, as we see shortly.

Case 3. This is the really important case, as it is not a limiting one. It consists of the generic condition in which the layer is far enough from the background to avoid a cast shadow, and illuminations falling on the layer and the background can have different intensities. If the layer is under low illumination, only a small amount of light will be added to the filtered portion of background luminance. If the layer is under high illumination, a proportionally large amount will be added. Note that this implies that a huge amount of light can be added as a common component to the filtered light. This component is critical when watching television under daylight, for instance, and is known as "veiling luminance" (Gilchrist & Jacobsen, 1983). In principle, when the assumption of homogeneous illumination is removed, the additive component of the mixture generated by the transparent layer can have any positive value.

4. THE EPISCOTISTER LUMINANCE MODEL

It is evident from the preceding discussion that the episcotister reflectance model meets both empirical and theoretical difficulties. These difficulties are connected largely with the upper boundary of the solution space. Apart from troublesome empirical evidence, one is also reluctant to accept a model that predicts that people cannot see layers under conditions of uneven illumination. Of course, people can do so.

A luminance version of the episcotister model (according to Case 3, section 3.2.) emphasizes the power of the scission hypothesis, and it also has higher ecological validity. I suggest it is applicable not only to optic arrays generated by classical pictorial displays, in which the stimulus is a mosaic of gray papers viewed in a normally light environment, but also to everyday vision, in which lightness constancy cannot be taken for granted.

Consider now *ABPQ* luminances and the simple hypothesis that *P* and *Q* are partitioned into a common component *F* and individual components proportional to *A* and *B*. In analogy with Metelli's formulas, equations for *P* and *Q* and solutions for partitioning parameters are as follows:

5. ACHROMATIC TRANSPARENCY

$$P = tA + F \tag{9}$$

$$Q = tB + F \tag{10}$$

$$t = \frac{P - Q}{A - B} \tag{11}$$

$$F = \frac{AQ - BP}{A - B} \tag{12}$$

Fractioning factor. Inspection of Equation 11 confirms that the fractioning factor t is a dimensionless number. Because t is a ratio between two luminance differences, it is invariant with respect to either absolute illumination and layer-to-background relative illumination. In ecological optics, t constitutes information specifying transmittance, a distal property. It is analogous to the ratio of two adjacent luminances, which is also invariant to changing illumination. However, it has the additional advantage that, whereas the latter specifies nothing more than *relative* reflectance, t specifies *absolute* transmittance. Again, t represents layer transmittance (i.e., a proportion): Hence, partitioning in the PQ region only occurs if t belongs to the $[0,1]$ domain.

Additive component. Inspection of Equation 12 indicates that F is not a number but a luminance. Furthermore, because it is a physical magnitude, the additive component F can take on any positive value. In the reflectance version of the episcotister model, where all variables are proportions, the qualitative difference between the fractioning factor (t) and the common component (f in the reflectance version and F in the luminance version) tends to be obscured. The physical meaning of F is complex. To understand it, let us see how the model is affected by the possible difference between background illumination I and layer illumination hI. Layer illumination is expressed as background illumination weighted by a factor h, which specifies the ratio of layer-to-background illumination. Layer luminances P and Q are a function of background reflectances a and b, layer transmittance t, layer material reflectance r, background illumination I, and layer-to-background illumination ratio h.

$$A = aI \tag{13}$$

$$B = bI \tag{14}$$

$$P = taI + (1 - t)rhI \tag{15}$$

$$Q = tbI + (1 - t)rhI \tag{16}$$

Analogously to the distinction in section 3, the light that is reflected by a layer has to be distinguished in terms of effective luminance F and material luminance L. The following relations hold:

$$F = (1-t)rhI \quad (17)$$

$$F = (1-t)L \quad (18)$$

$$L = rhI \quad (19)$$

$$L = \frac{AQ - BP}{(A+Q) - (B+P)} \quad (20)$$

The quantity L is a luminance, derivable from $APQB$ luminances. It is

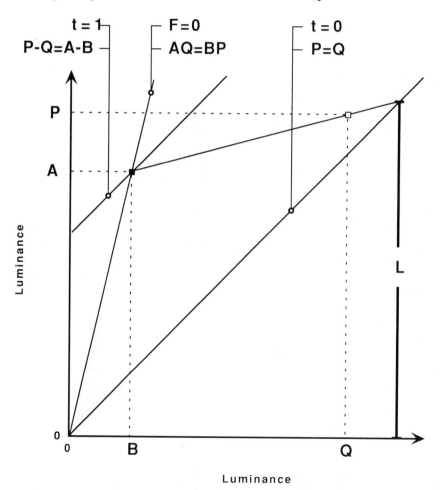

FIG. 5.12. Graphical representation of the four-intensity pattern in the luminance space.

the illumination-variant component of F. This is because the (1 − t) component obviously is illumination-invariant. The term to the right of the equality sign in Equation 20 is formally equivalent to its counterpart in Equation 5. Both yield a ratio between the difference of cross-products and the difference of cross-sums (*cross-* denotes the result of operations upon nonadjacent regions). However, its interpretation here is different. Material luminance L contains three unknowns: material reflectance, r; the ratio of layer-to-background illumination, h; and background illumination, I. If the four luminances provide the only set of independent variables, one cannot derive r without additional assumptions. This should not surprise us because the same is locally true of each background reflectance. Unless illumination is specified, reflectances cannot be recovered from luminances. To develop this point, let us modify the Remondino diagram and adapt it to the luminance domain (Fig. 5.12).

The general conventions described previously for the reflectance space are also used to represent the *APQB* pattern in the luminance space. Given the (B,A) point, the solution region is an open trapezoid defined by three boundaries: $t = 0$, $t = 1$, and $F = 0$ (coincident with $r = 0$). Following Beck et al. (1984) and Brill (1984), these boundaries can be alternatively expressed by a set of constraints upon partitioning of P and Q. Under the assumption that $A > B$, these are the constraints: $P > = Q$; $(A − B) > = (P − Q)$; $AQ > = BP$. Given both the (B,A) and (P,Q) points, then t and F are immediately given. Like in the reflectance space, t here is a ratio between two segments, but the segments now are luminance differences. On the other hand, F is not a ratio but a luminance on the same scale used for A, B, P, and Q.

Now, connect (B,A) and (P,Q) points by a straight line and find its intersection with the main diagonal $y = x$. The vertical (or horizontal) coordinate of this intersection is L, the material luminance. At this point an arbitrary option is needed. Figure 5.13 illustrates the three possible settings of background illumination I with respect to L, and their most plausible interpretation:

$I = L$, $rh = 1$: Background illumination has the maximum value compatible with both maximum layer reflectance ($r = 1$) and homogeneous illumination ($h = 1$);

$I > L$, $rh < 1$: Layer reflectance has a valid value ($r < 1$) and illumination is homogeneous ($h = 1$); this option is virtually identical to Metelli's reflectance model;

$I < L$, $rh > 1$: according to Case 3 (section 3.2.) and Partitioning 4 (Fig. 5.6), layer reflectance is maximum ($r = 1$) and layer illumination is higher than background illumination ($h > 1$).

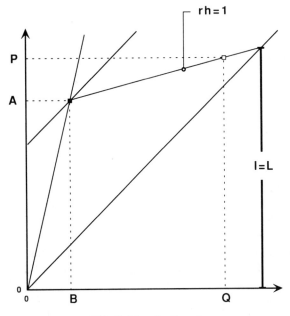

FIG. 5.13. Continued.

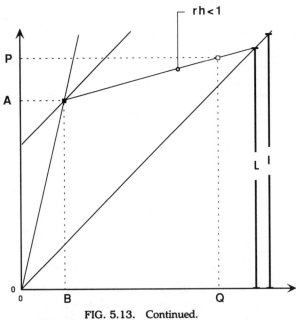

FIG. 5.13. Continued.

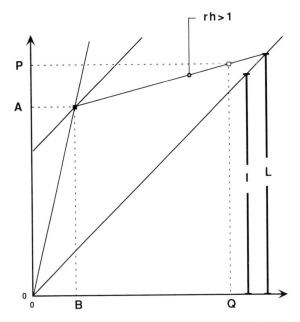

FIG. 5.13. Three possible settings of background illumination I with respect to L.

Only in Option 3 one is forced to have two different illuminations. Because of the rh trade-off, however, one could arbitrarily increase layer illumination ($h > 1$) in Options 1 and 2 as well. Furthermore, in Option 2 layer illumination can be decreased ($h > 1$) if r is proportionally increased. Although formally correct, these possibilities violate the minimum principle; therefore, $h = 1$ in Option 2 and $r = 1$ in Option 3 seem reasonable default values.

5. THE FILTER REFLECTANCE MODEL

The filter model for predicting transparency percepts was originally introduced by Beck et al. (1984). Although Beck et al. did not explicitly state this, the model appears to apply to only a specific illumination condition; namely, Case 1 of section 3.2. Like Metelli, Beck et al. also used reflectances as independent variables.

Algebraically, the filter reflectance model appears to be far more complex than the simple episcotister reflectance model. It is more complex because it takes into account the complex mixture produced by a layer with transmittance t and effective reflectance f, located in front of an opaque surface with reflectance a, whose occluded portion is illuminated only through the layer. This is illustrated in Fig. 5.14, where different components that determine virtual reflectance p are analyzed.

Consider again a pq pair as we did in the episcotister reflectance model. Parallel equations for determining the pq pair as a function of t, f, and background reflectances ab are given in the filter reflectance model:[6]

$$p = t^2 \frac{1}{1-fa} a + f \qquad (21)$$

$$q = t^2 \frac{1}{1-fb} b + f \qquad (22)$$

Two assumptions about physical conditions lead to these equations for virtual reflectances p and q:

1. The occluded background is assumed to be indirectly illuminated. This may happen either because illumination is directional or because the layer is in contact with the background (illumination's direction irrelevant). However, a cast shadow is produced. Therefore, p does not contain the ta component, which makes sense only if the background is directly illuminated.

2. Sandwich reflections are assumed to be present and are taken into account. Light can bounce between layer and background, and the amount transmitted to an observer's viewpoint becomes smaller and smaller after each partial reflection. Infinite filtering produces the series $(t^2 a + t^2 a^2 f + t^2 a^3 f^2 + \ldots)$ converging at $t^2 a / (1 - fa)$.

There are remarkable structural similarities between episcotister equations 2 and 3 and filter equations 21 and 22. Both the episcotister model and the filter model contain an additive term, and both have a factor that fractions background reflectances. However, in the filter model the fractioning factor is not simply layer transmittance; instead it depends on both the additive term and the specific background reflectance. Given that $f = (1 - t)r$ holds here too, solutions for transmittance t, effective reflectance f, and material reflectance r are as follows:

$$t = \frac{\sqrt{(a-b)(p-q)(1-aq)(1-bp)}}{a(1-bp) - b(1-aq)} \qquad (23)$$

---

[6] Filter equations are rewritten in a form that looks simpler (at least to me) and facilitates the comparison with episcotister equations. However, they are equivalent to those provided by Beck et al. (1984, p. 415) and Brill (1984). Solutions for layer parameters are also rewritten to be more meaningful, also with respect to the successive luminance version.

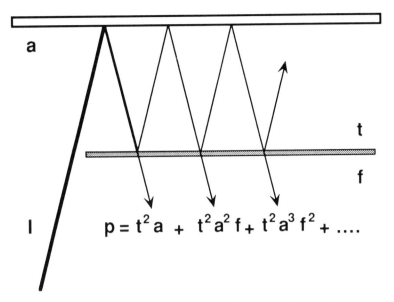

FIG. 5.14. Virtual reflectance p according to the filter model.

$$f = \frac{aq - bp}{a(1-bp) - b(1-aq)} \qquad (24)$$

$$r = \frac{aq - bp}{a(1-bp) - b(1-aq) - \sqrt{(a-b)(p-q)(1-aq)(1-bp)}} \qquad (25)$$

According to Beck et al. (1984), both t and f (and therefore r) belong to the [0,1] domain. Here, the upper limit of the additive component is not arbitrary as it is in the Metelli model. However, this is a clear consequence of the fact that the environmental arrangement addressed always involves homogeneous illumination (on both layer and nonoccluded background); that is, physically the filter model is correct in conditions described by Case 1 (section 3.2.), in which the episcotister model is not. But it is incorrect in all other conditions. These considerations reflect restrictions on the filter model as a physical model. However, if we consider it simply as an algebraic model, it does provide valid solutions to the partitioning problem in the reflectance domain, but it does so in a mathematically cumbersome fashion.

6. THE FILTER LUMINANCE MODEL

The exclusive reliance on reflectances instead of luminances is even less justifiable in the filter model than in the episcotister's. This is because none of the layer parameters is invariant to illumination changes. The lack of invariance becomes apparent in the luminance version of the filter model.

$$A = aI \tag{13}$$

$$B = bI \tag{14}$$

$$P = t^2 \frac{I^2}{I^2 - FA} A + F \tag{26}$$

$$Q = t^2 \frac{I^2}{I^2 - FB} B + F \tag{27}$$

$$t = \frac{\sqrt{(A-B)(P-Q)(I^2-AQ)(I^2-BP)}}{A(I^2-BP) - B(I^2-AQ)} \tag{28}$$

$$F = \frac{I^2(AQ-BP)}{A(I^2-BP) - B(I^2-AQ)} \tag{29}$$

Therefore, because $F = fI = (1 - t)rI$:

$$r = \frac{I(AQ-BP)}{A(I^2-BP) - B(I^2-AQ) - \sqrt{(A-B)(P-Q)(I^2-AQ)(I^2-BP)}} \tag{30}$$

Of course, f simply is the F/I ratio. Inspection of Solutions 28 and 29 reveals that t is directly related to the illumination factor I, whereas F is inversely related to it. More interestingly, filter solutions converge toward those from the episcotister model as I increases. This can be seen by comparing Equations 9 and 26: Notice that as I grows, the fraction of Equation 26 approaches unity, and P luminance simply becomes $t^2A + F$.

Therefore, when a large illumination is postulated, the only difference between the two models is the meaning attributed to the fractioning factor of the partitioning function. In the episcotister model (i.e., when the occluded background is directly illuminated), this component is simply transmittance t, whereas in the filter model (i.e., when the occluded background is illuminated only through the layer), this component is squared transmittance t^2. One can reach the same conclusion by looking at Equations 28 and 29: As I grows, solutions for t^2 and F become identical to solutions for t and F in the episcotister luminance model (Equations 9 and 10).

Figure 5.15 illustrates this convergence when $APQB$ luminances are equal to 10, 12, 9, 4 (in arbitrary luminance units), and the illumination factor I is increased from 12 (equal to P) up to 72 (p reflects only one-sixth of the incident light). In the episcotister luminance model, the invariant

5. ACHROMATIC TRANSPARENCY

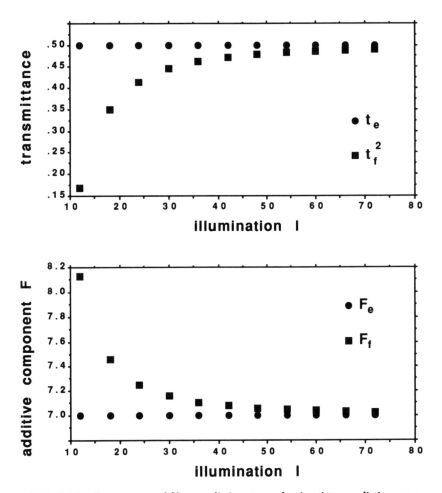

FIG. 5.15. Convergence of filter predictions toward episcotister predictions as a function of the illumination factor.

value of t is .5 and the invariant value of F is 7. In the filter luminance model, t^2 varies from .17 to the asymptote value .5, and F varies from 8.13 to the asymptote value 7.

The filter model is physically correct in several everyday occurrences of transparency. For instance, it applies to thin sheets of paper superposed on a table as well as to a veil covering a face. However, it does not strongly recommend itself as a predictive tool in perceptual theory. Whereas its good features are present in the episcotister model in a more general form, it cannot be used to solve the partitioning problem unless the illumination factor is specified. The following section regards a comparison of theoretical predictions of the two models under a specific illumination.

7. A COMPARISON OF EPISCOTISTER AND FILTER PREDICTIONS

After comparing solutions offered by episcotister and filter reflectance models using a computer search, Beck et al. (1984) claimed that the solution regions for the two models are "closely related" (p. 416). However, Beck et al. were apparently equating material reflectance r derived from the episcotister model and effective reflectance f derived from the filter model. Figure 5.16 visualizes differences between the two models when background reflectances are $a = .6$ and $b = .2$.

Boundaries corresponding to $t = 0$ and $r = 0$ are the same in both models. In these two cases, constraints are the same for both models. Line $t = 0$ is $y = x$; both Equations 3 and 22 become null when $p = q$. Line $r = 0$ is $y = (a/b) x$; both Equations 5 and 25 become null when $aq = bp$.

Although actually a curve, Beck's $t = 1$ boundary is very closely approximated by a straight line with slope 1.326 (from Equation 23). Such a value is specific to the ab pair chosen in Fig. 5.16. The slope of this boundary varies as a direct function of both reflectance contrast (high a/b ratio leads to high slope) and absolute reflectance level (keeping a/b constant, high ab values lead to high slope). This is a major disadvantage with respect to the episcotister model, in which the slope of all iso-t lines is invariant to background changes. Beck's $r = 1$ boundary is a monotonic curve through (b,a) and $(1,1)$ points, lower than Metelli's one: $y = .57 + .08 x + .34 x^2$. This means that the filter model is more restrictive than the episcotister model.

Beck's solution region is smaller than Metelli's. For instance, consider solutions for two pq pairs belonging to the mismatch region bounded by Metelli's $r = 1$ and Beck's $r = 1$, always relative to the (.6, .2) background pair. When $p = .7$ and $q = .45$, Beck's $t = .68$ and $r = 1.09$ (invalid), whereas Metelli's $t = .62$ and $r = .87$. When $p = .85$ and $q = .75$, Beck's $t = .34$ and $r = 1.11$ (invalid), whereas Metelli's $t = .25$ and $r = .94$.

A direct comparison of episcotister versus filter predictions is provided in Fig. 5.17 and 5.18. Disks represent the absolute discrepancy between episcotister and filter predictions when the A/B ratio is equal to one fourth. Parameters of the filter model were evaluated under the assumption that the highest luminance in the field sets the illumination factor $I = A$. Inspection of the two graphs indicates that discrepancies can be rather large for certain $ABPQ$ combinations.

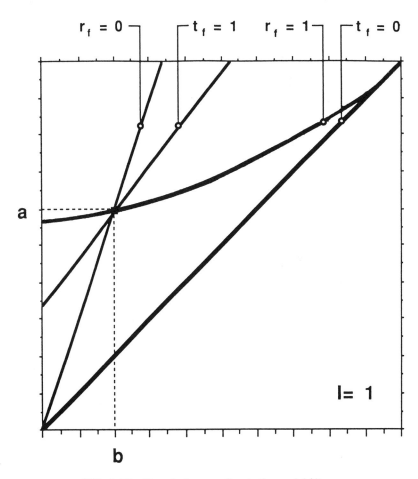

FIG. 5.16. Boundaries according to the model filter.

8. AN EVALUATION OF THE SCISSION HYPOTHESIS

Generally, two kinds of tasks have been used to test Metelli's theory:

1. Absolute judgment task: Given a single four-intensity pattern, the subject must report whether it appears transparent or not (yes–no judgment), or evaluate the apparent degree of transparency (rating, magnitude estimation).

2. Relative judgment task: Given two patterns with the same *AB* background pair but different *PQ* layer pairs, the subject must tell which layer is more transparent.

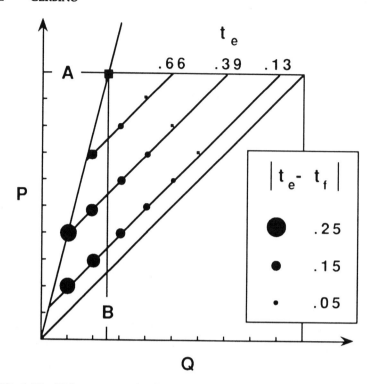

FIG. 5.17. Disks represent the absolute discrepancy of the t value derived from Metelli's and Beck's models when $A/B = 1/4$, under the assumption that $I = A$.

Both tasks are contaminated by a serious methodological problem. The problem is related to two difficulties: The first lies in the ill-defined meaning of the term *transparency* in everyday language; the second refers to the observer's ability to attend to both relevant properties of the layer, reflectance and transmittance, and judge accordingly to both.

The idea of partitioning into common and individual components might lead to the conclusion that the layer's perceived color corresponds to effective reflectance and not to material reflectance. Metelli proposed that the eye picks up material reflectance r, which makes a lot of sense. The same effective reflectance f may result from either a dense layer of low reflectance or a tenuous layer of high reflectance, keeping illumination constant. Because normally we do not confuse a dark thick veil with a light thin veil, the perceived color of the layer should be a function of material reflectance r, not of effective reflectance f. Whereas this aspect of Metelli's theory is very reasonable and perfectly consistent with the underlying physical model, it has its own weakness. For totally dense surfaces, the distinction between effective and material reflectances breaks down; when constancy occurs, achromatic color of an opaque surface is a function of reflectance, along the black–white continuum. Can we claim the same for transparent surfaces as well?

Wittgenstein (1953) observed that black certainly is a color of the kind red, green, yellow, and blue are, whereas white is not. He had in mind some observations with colored glasses. According to him, we can say that a piece of transparent glass is red, green, or black; but we can call it white only when it is opaque. This association of whiteness with opacity is a matter of common experience: Keeping density constant, we see better through a veil made of black fibers than through a veil made of white fibers. This fact cannot be easily understood if our theory predicts that a layer can be perceived as both maximally trasmittant (t near 0) and maximally "white" (r near 1). It seems that the black–white continuum used for opaque surface colors should not be used also for transparent colors.

According to these remarks, tasks involving explicit judgments of transparency should be avoided. On the one hand, these tasks require too much from the observer; on the other, they are not adequate to test the precise quantitative predictions derivable from an algebraic model.

Consider the analogy with lightness constancy in surface-color perception. One can ask subjects to rate the lightness of a single region sur-

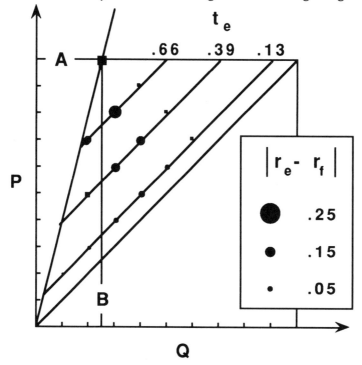

FIG. 5.18. Disks represent the absolute discrepancy of the r value derived from Metelli's and Beck's models when $A/B = 1/4$, under the assumption that $I = A$.

rounded by another, or to choose the lighter between two regions having the same surround (which is rather trivial only because lightness is unidimensional, whereas transparency is probably not). But, one can better ask subjects to make a lightness match of two different regions within different surrounds: This procedure leads to the discovery that lightness perception is controlled by luminance ratios (Jacobsen & Gilchrist, 1988b; Wallach, 1948, 1976).

The scission hypothesis embodied in the episcotister model (luminance version) corresponds to the ratio principle of adjacent luminances. In fact, it can explain a particular kind of object constancy (i.e., perceptual constancy of a layer superposed upon different backgrounds). Two identical layers on different two-color backgrounds generate completely different four–luminance patterns. Despite inequalities between corresponding luminances of each pattern ($A \neq A'$, $P \neq P'$, $Q \neq Q'$, $B \neq B'$), the episcotister luminance model predicts that the two layers look identical because t and F derived from $APQB$ are identical to t' and F' derived from $A'P'Q'B'$. The predicted matching is veridical and invariant to changing common illumination. According to the general notion of direct perception, intensity relationships described by the episcotister luminance model will optically specify layer constancy.

Brill (1984) correctly emphasized that only scale-invariant properties of the optic array can constitute information for transparency. Following this general idea, he claimed that constraints derived from the value of r (in the reflectance formulation) cannot be perceptually relevant, because this value is not scale-invariant. Such a claim is theoretically justified when absolute judgments are required. However, it is not valid for a matching task, which is the appropriate psychophysical operationalization of object constancy. As demonstrated in section 4, in the episcotister luminance model t is a dimensionless number (therefore, it is an absolute invariant), whereas F is scaled on the same continuum as A and B. In other words, F is invariant with respect to the illumination common to both layer and background. Therefore, F can control relational perception (i.e., matching). Things are completely different in the filter model, in which both t and F vary as a function of the common illumination.

A matching experiment was performed by Gerbino et al. (1990). They used 24 pairs of four-luminance patterns, generated by a computer on a high-resolution monitor. In each pair, the left pattern was the standard and the right pattern the comparison; the standard $APQB$ pattern and the comparison $A'B'$ background pair where kept constant, whereas the comparison $P'Q'$ pair could be adjusted. Subjects had to adjust $P'Q'$ luminances and match the two perceived layers. Adjustment was constrained along constant t, which means that subjects could change only the common additive component F (values derived from the luminance version of the episcotister model). No transparency judgment was required. Eight A/B ratios (ranging from 2.5 to virtual infinity) and three t values (.25, .50, .75) were used. Layer luminances were always higher than background luminances; that is, the left maximum luminance was P and the right maximum luminance P'. This means that the whole experiment was conducted in a region that is "dangerous" according to the reflectance version of the episcotister model. In fact, if patterns were perceptually scaled taking the highest background luminance (for instance, A) as the illumination factor, then an unadmissible value $r > 1$ would have resulted. Therefore, the experiment tested both the superiority of luminance versus reflectance versions and the suggestion that matching (i.e., layer constancy) is possible independently from the specific interpretation given to the common additive component F (see options described in section 4, p. 550, row 1).

Adjustment data obtained by Gerbino et al. (1990) fit predictions of the episcotister scission model better than alternative models not based on scission.

ACKNOWLEDGMENT

Part of the work reported in this chapter has been supported by the CNR grant no. 89.00515.67 to the unit "Local Vision" of the Robotics Project.

6
Color Constancy: Arguments for a Vector Model for the Perception of Illumination, Color, and Depth

Sten Sture Bergström
University of Umeå

The author has earlier made attempts to apply Johansson's perceptual vector analysis to the area of color and illumination (Bergström, 1977a, 1982; Bergström, Gustafsson, & Putaansuu, 1984).

It was assumed in this application that light reflected by illuminated objects is automatically analyzed into common and relative (specific) components. The visual system was assumed to be able to distinguish between the illumination component (called *ilco*) and reflectance components (*recos*) in the proximal stimulus. And it was assumed that this distinction is possible thanks to the common component characteristic of the illumination. This assumption (the commonality assumption) means that the visual system can discriminate between the retinal projection of an illumination border and that of a reflectance border and between the retinal projection of a shadow and that of a darker color because illumination has this characteristic of being a common component.

It was also assumed that the modulation of illumination in space and/ or time corresponds to the perceived 3-D layout of the scene and to motion in depth within the scene.

Empirical support to the commonality assumption is given from both the literature (from Katz, 1911, to Metelli, 1983) and from experimental studies performed in my own laboratory. It is shown that our model (the two preceding assumptions plus a minimum principle as to the number of light sources) can be applied to the results of a number of color (lightness) constancy studies (both constancy and nonconstancy results). It is applied to Land's Mondrian demonstrations, to Gilchrist's two classes of experiments, one class giving a shift of perceived lightness with a shift of viewing conditions, the other giving no such shift. And, finally, it is also discussed in relation to Metelli's demonstration of a hole in a screen appearing like a surface color and to color rendering. The problem of reflectance edges versus illumination edges is also discussed within the same theoretical frame of reference (cf. the discussion in section 3.2 and Gustafsson, 1987; Gustafsson, Bergström, & Jakobsson, 1987; Bergström, Gustafsson, & Jakobsson, 1993).

1. PERCEPTUAL CONSTANCY

Perceptual constancies are phenomena in which a variation of the proximal stimulus does not give the expected perceptual variation. A shrinking retinal representation does not appear as a shrinking distal object but as a motion of the object away from the observer. A change of the geometrical form of the retinal representation of a swinging door does not appear as the door changing form but as a change of its orientation (slant) relative to the position of my eye. And a change of the spectral composition of the light reflected from an object does not appear as a change of its color but as a change of the illumination, a.s.o. The list could be made much longer.

The perceptual constancies caught the attention of perceptionists long before they were called psychologists, and the literature on the topic is enormous. (For a good historical review, see Epstein, 1977, and for a review of the classical experiments, see Woodworth & Schlosberg, 1955, chap. 15 and 16.)

To empiricists, the constancies became nice examples of perceptual phenomena, which had to be explained by earlier experience. To nativists, the search for stimulus invariances corresponding to the perceptual constancies became a challenge. The constancies seem to be necessary for our veridical perception, and they have certainly played an important role for our survival—and still do.

Our first question concerning the constancies should probably not aim at an explanation of the phenomena. Instead we should ask ourselves why we had the wrong expectations as to what will appear to vary when the proximal stimulus varies. The reason, of course, is that our theories are inadequate. They will probably continue to be poor, but putting the right questions forth may help to improve them.

Rather than relying on effects of early experience, it seems currently a more fruitful strategy to assume proximal invariances corresponding to the perceived constancies and to search for alternative descriptions of the proximal stimulus. This is not the same as assuming perception to be uninfluenced by experience. But Hering (1964) argued as early as 1874 that some degree of constancy must be innate because it is a precondition of learning to identify objects (p. 21).

Johansson (1977) argued that a shift of stimulus description from the absolute Euclidean geometry to a projective geometry assuming invariant relations at the proximal level solves the problems of size and form constancy. In central projection the shrinking image is information about motion in depth, and systematic variation of angles is information about the rotation of orthogonal objects in space a.s.o. Reasonably enough, the visual system seems to be designed to take advantage of the information afforded by the optical projection system that is a part of it.

2. COLOR CONSTANCY AND A VECTOR MODEL

The problem of color constancy, on the other hand, has not been brought to such an elegant solution as the size and form constancies. But certainly scientists have been looking for invariant stimulus relations to replace the classical dominant wavelength explanation of color perception. Helson's studies on adaptation (Helson, 1938; Helson & Jeffers, 1940; Helson & Michels, 1948), Judd's (1940, 1960) equations, Wallach's (1948) luminance ratio model, and Land and McCann's (1971) Retinex theory can be mentioned as examples.

A parallel to the shrinking and growing square (Johansson, 1964) but within the field of color constancy is a luminance gradient perceived as a three-dimensional object illuminated from a certain direction (Bergström, 1977b; Mach, 1959; see Fig. 6.1).

In the luminance gradient, like in Johansson's shrinking and growing square, there is a perceptual constancy "at the cost of" a nonveridical three-dimensionality. Other such examples are the shadow-cast experiments by Metzger (1934a, 1934b) and by Wallach and O'Connel (1953). An interesting example of speed constancy at the cost of a nonveridical three-dimensionality is reported by Holmgren (1974), who presented moving dot patterns with spatial velocity gradients. Still other examples are given by Bergström (1980, 1982). An illumination border across a picture makes it appear folded in depth. Varying the two illumination levels (of the two halves of the picture) in counterphase makes the two halves appear to swing in depth like doors (Fig. 6.2). However trivial these nonveridicalities may appear, they have something very important to tell us about the function of the visual system, about what will stay constant and what will vary as a function of varying the proximal stimulus (i.e., about the hierarchy of constancies).

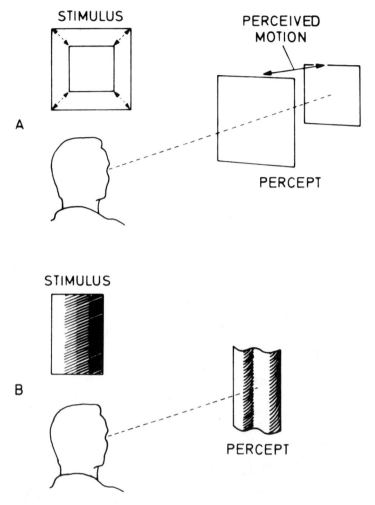

FIG. 6.1. The "shrinking and growing square"-stimulus and the corresponding perception of motion in depth of a rigid object (A). The luminance gradient as stimulus perceived as a three-dimensional object illuminated from one side (B) (From Bergström, 1977b).

2.1. The Main Assumptions of our Model

In a couple of earlier papers (Bergström, 1977b, 1982) I tried to apply to reflected light Johansson's (1950, 1964) analysis of motion into common and relative components. It was assumed that light reflected from illuminated objects is automatically analyzed into common and relative components. The visual system was assumed to be able to distinguish between

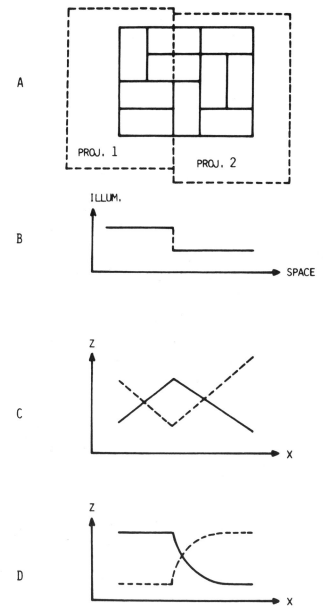

FIG. 6.2. The "Mondrian" picture illuminated by two projectors (A). The cyclic variation of the illumination from the two projectors is illustrated by Diagrams B and C. Diagram D illustrates the corresponding reported movement to and fro in depth (From Bergström, 1982).

the illumination component (the *ilco*) and reflectance components (*recos*) in the proximal stimulus. And it was assumed that this distinction is possible thanks to the common component characteristic of the illumination. Thus, commonality is assumed to be the distinctive feature that makes color constancy possible. Just as the visual system immediately extracts a common motion component from a pattern of moving objects, so it extracts a common component of illumination from the pattern of reflected light. The assumptions mean that the visual system can discriminate the retinal projection of an illumination border from that of a reflectance border and the retinal projection of a shadow from that of a darker color because the illumination has this characteristic of being a common component.

A second assumption is that modulation of the illumination in space and/or time informs us about the 3-D layout of the distal scene and about motion in depth within that scene.

One of the characteristics of illumination is, of course, its chromatic color. This is illustrated in the marvelous color constancy demonstrations by Land and McCann (1971) in their Mondrian experiments. A "Mondrian" is a stimulus display that consists of a number of rectangular color samples arranged to resemble some of Piet Mondrian's paintings. Land and McCann demonstrated a high degree of color constancy even under drastic spectral variation of the illumination of such a display. And a common chromatic component is what artists apply when painting a motif as if it were in a certain illumination.

2.2. Retinal Illumination Analyzed

Now, as an illustration, let us consider Land and McCann's demonstration using an achromatic Mondrian (i.e., a Mondrian consisting of different shades of gray only). This Mondrian picture was illuminated from below in such a way that a gray field at the bottom got the same mean luminance as a white field at the top. Still, of course, the gray field looked gray and the white field looked white. According to Land and McCann (1971), this constancy is the result of the low sensitivity of the visual system to gradual luminance variations in space. According to our model, it is exactly the other way around. The visual system is not only sensitive to the gradual variation of luminance, but it also recognizes it as a common component (i.e., as illumination). The Mondrian is outlined in Fig. 6.3 and so is the luminance profile along one of the possible paths across the picture, which plays an important role in the Land and McCann theory. This luminance profile (one aspect of the distal stimulus) has its counterpart in the projection into the eye (i. e., in the proximal stimulus; see Fig. 6.4).

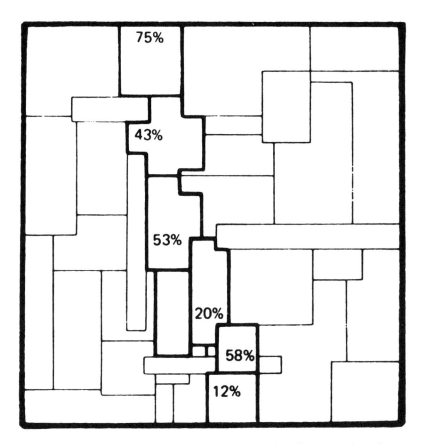

FIG. 6.3. A sketch of the achromatic Mondrian with the reflectance values along a track from the gray at its bottom to the white at its top (From Land & McCann, 1971).

The proximal stimulus profile is assumed to be automatically analyzed into an illumination component (*ilco*) and a reflectance component (*reco*). The ilco is proportional to the common change of luminance across the whole picture. The residual component is proportional to the different reflectances along the path.

The ilco and recos are assumed components corresponding to the perceived illumination and the perceived reflectances, but they are the results of our assumed analysis—not percepts.

It should be stressed here that the proposed analyses are not simple subtractions, because luminance is a product of reflectance and illumination, not a sum. Technically, this problem can be solved simply by using the logarithms of the luminance values instead of the values themselves.

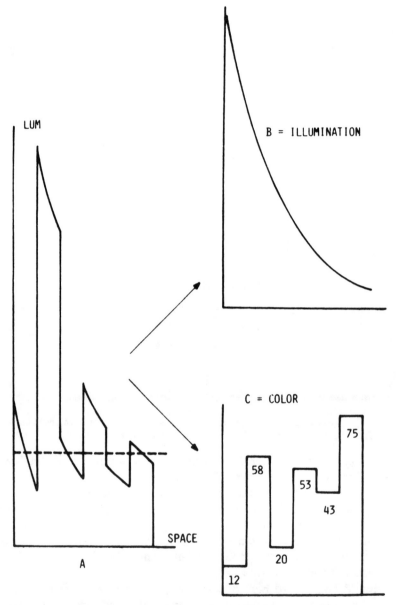

FIG. 6.4. A sketch of the luminance profile along the track of Fig. 6.3 (A) and the analysis into a common illumination (B) and relative lightness components (C). The figures denote reflectances from the gray (12%) to the white (75%) sample (From Bergström, 1977b).

3. EMPIRICAL SUPPORT TO THE COMMONALITY ASSUMPTION

The assumption that variations of illumination can be distinguished from variations of reflectance thanks to the commonality characteristic of illumination is the key assumption of our model. We hope it will solve the problem of color constancy.

The capacity of the visual system to extract common motion components has been demonstrated in a very compelling way originally by Johansson (1950, 1964) and later by other members of the Uppsala group (Börjesson & von Hofsten, 1972, 1973, 1975; Marmolin, 1973a, 1973b, 1977; Runesson & Frykholm, 1983), and also in other laboratories (e.g., Cutting, 1978).

To support our assumption we need empirical evidence that conditions revealing the common component characteristics of illumination will also favor color constancy, whereas conditions concealing those characteristics will jeopardize the constancy. As a matter of fact there is a lot of such evidence in the literature.

3.1. One Field Versus Two or More Fields in the Same Illumination

Because two is the lowest number of fields to have an illumination in common, there should be quite a difference as to color constancy between one single homogeneous illuminated field and two or more fields in the same illumination.

One of the first systematic studies on color constancy was performed by Katz (1911), who had his observers match color wheels, one of which was in direct illumination and the other in shadow. It appeared that his observers adjusted to almost exactly the same gray in shadow as in direct illumination. To get a measure of the degree of constancy, Katz needed a matching without any constancy at all. His trick to get a "zero constancy" condition was to use a reduction screen (i.e., a screen having a peephole permitting the observer to see only the target (the color wheel) and no immediate surrounding).

Gelb (1929) made a very nice demonstration using a concealed lantern to illuminate only a black disk in a room, the rest of which was very dimly illuminated. The black disk appeared white. As soon as a piece of real white paper was inserted within the beam of the concealed lantern, the apparent white disk turned black. When the small white strip was removed, the disk immediately flipped back into white again, despite the observer's knowledge of the extra illumination. The effect is known as the

Gelb effect. Its independence of the observer's conscious knowledge is significant because it indicates the automaticity of the analysis of the proximal stimulus.

Kardos' (1929, 1934) experiments using a concealed shadow-caster were analogous to Gelb's aforementioned experiment.

Hsia (1943) compared the constancy of test surfaces in a common room illumination with that of separately illuminated surfaces. He found no difference. The separate sources of light in Hsia's experiment illuminated *a test surface and its immediate surround*. Apparently, the visual system had no problem identifying a number of different illuminations, each of which was common to just one test surface and its immediate surround.

Land and McCann (1971) reported a very compelling demonstration of color constancy using a complex chromatic picture (a Mondrian) and a three-projector set-up to manipulate its illumination. They stressed the importance of the complexity of the picture to get constancy, but Bergström (1977a), using basically the same arrangement and a peephole technique, reported that two colored fields (or one field and its nearest surround) are necessary and sufficient to give color constancy under those conditions.

All these results support the assumption of a common chromatic component of illumination.

3.2. Simulated Attached Shadows

Bergström (1980, 1982) and Bergström, Gustafsson, and Putaansuu (1980, 1984) reported experiments where test surfaces were illuminated by a projector using a slide with a square-wave grating. Gratings of varying contrast (modulation depths) were superimposed on two different test surfaces: a simplified chromatic Mondrian and a homogeneous gray surface.

The observers were instructed to report whether they perceived the test surface as a flat (striped) surface or as a three-dimensional (pleated) object. Three-dimensional responses were considered to indicate color constancy as the dark phases of the grating simulated attached shadows. Two-dimensional responses are not conclusive as to color constancy because the observers did not report whether the darker stripes appeared as shadows or as darker colors. The 3-D responses in this ambiguous situation had the advantage of indicating color constancy without any instruction about colors or shadows.

If we apply the same type of analysis to these stimuli as we did to the achromatic Mondrian, we find alternative possible perceptual outcomes, which is reasonable as the situation really is ambiguous. Figure 6.5, 6.6,

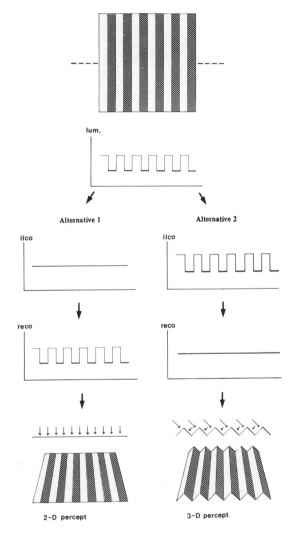

FIG. 6.5. Two alternative analyses of the luminance profile (or rather retinal illuminance profile) of a homogeneous gray display under the "grating llumination." Both alternatives seem to be equally simple, and 2D and 3D equally probable. Ambiguity is predicted. The proximal luminance profile is called *lum*, and *ilco* and *reco* mean the illumination and reflectance components, respectively (modified from Bergström et al., 1984; reprinted from *Perception*, 1984, 13, 129–140, Pion Limited, London).

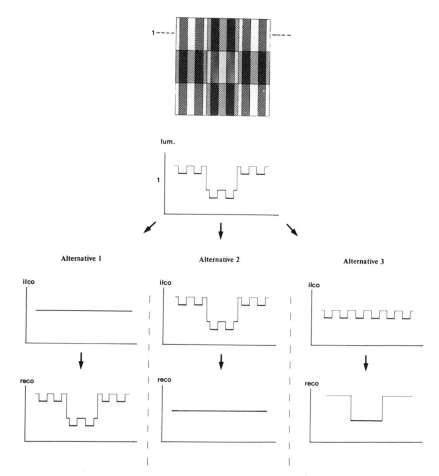

FIG. 6.6. Three alternative analyses of one profile of the Mondrian. Which alternative is the simplest one?

and 6.7 illustrate the analysis of the homogeneous gray and the Mondrian displays with the superimposed square-wave grating.

For the homogeneous gray we get two alternatives, one giving a three-dimensional, the other a two-dimensional percept. Thus, ambiguity will be predicted.

In the Mondrian condition we have illustrated three alternatives. They look very much the same as to simplicity. But as soon as we consider the two different rows of squares in the Mondrian, we find that the repetitive variation is common to both rows (or, as a matter of fact, to the whole display as it is symmetrical).

Using intuitive criteria of simplicity, it seems that Alternative 3 is

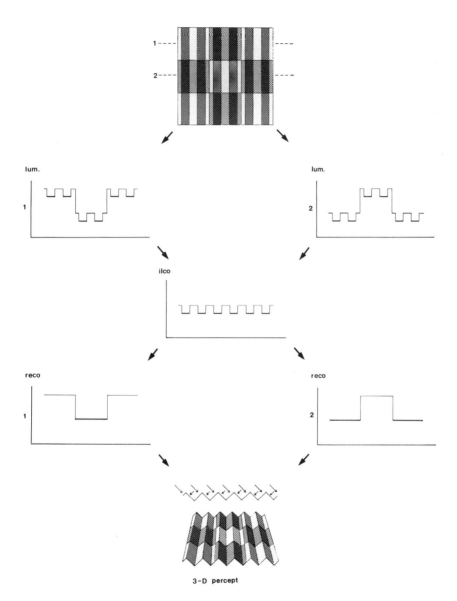

FIG. 6.7. An analysis of two profiles of the Mondrian makes the alternative assuming a common modulation of the illumination simpler than other alternatives (from Bergström et al., 1984; reprinted from *Perception*, 1984, *13*, 129–140, Pion Limited, London).

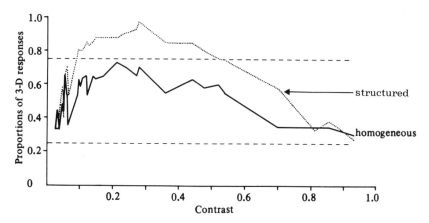

FIG. 6.8. Proportions of 3-D responses to the homogeneous and the structured (Mondrian) displays. The 0.25 and 0.75 threshold values are marked in the diagram. (From Bergström et al., 1984, Experiment 1; reprinted from *Perception*, 1984, 13, 129–140, Pion Limited, London).

simpler than the others. This means that we will predict three-dimensional percepts to dominate for the Mondrian condition, whereas the homogeneous gray will be ambiguous.

Our predictions are based on still another assumption, however: that modulation of illumination informs us about the three-dimensional layout of the distal scene.

It must be stressed here that our prediction is contrary to what would be predicted from traditional theories on depth perception. According to them, the Mondrian would appear flat because the cues for depth are absent, the borders between fields would not converge and diverge in harmony with the grating; they would still be straight, parallel, and uninfluenced by the modulation of the illumination. The homogeneous gray surface, on the other hand, has no such cues to flatness and could be predicted to be more vulnerable to manipulations of the illumination.

The results are definitely in line with our predictions as can be seen in Fig. 6.8.

The frequencies of 3-D responses are significantly higher for the Mondrian than for the homogeneous gray test surface.

Although being paradoxical from the standpoint of traditional cue theories, the results strongly support the basic assumption of our model. The structure of the Mondrian (the very same structure that gives cues to its flatness) reveals the common component characteristic of the illumination (and of the shadow), whereas no such information is present in the homogeneous gray surface (cf. the lack of color constancy in Gelb's black disk referred to earlier).

The observers in the Bergström et al. experiment saw the test surfaces evenly illuminated in the pauses between the exposures of the grating, which means that they were aware of the real conditions. This supports the assumption of automaticity as the Gelb demonstration does.

3.3. One Rather Than Two Sources of Light

In assuming a common component of illumination, we have not yet said anything about what that is.

In our daily life we are used to more than one source of light at a time. As long as they are all about the same color, they add up to a total illumination of the room or of our visual field, and there is normally no problem with color constancy. (There will be a number of shadows, however, and with different color temperatures of the sources these shadows will be different colors.) The common component extracted is the room illumination, whatever it might be, with all its modulations and complexity. The optical projection into the eye from any one object in the visual field informs us about this complex room illumination and so does our peripheral vision.

In our laboratories we often create situations that are ambiguous as to illumination. Metelli (1983) demonstrated how a hole in a screen can be perceived as a surface color. He used a black screen with two rectangular fields, one white, the other gray. A third field of the same size is in reality an aperture in the screen into a case containing a lamp giving a blue illumination. The observer is instructed to touch the three rectangular fields using a stick. To his or her surprise, the stick goes through the hole when trying to touch the blue rectangle.

First question: Why does the hole look like a surface?

Second question: Why do not all the rectangles look like holes?

Answers:

1. The hole has a luminance within a critical 30:1 ratio compared to its total surrounding (see following). This allows us to perceive it as a surface color. For something to look like an aperture, two illuminations have to be assumed, one of which does not influence the rest of the visual field (i.e., it would not be a common component). Apparently, we see a light as an illuminant only if it is common to more than one object or if it exceeds the 30:1 ratio. In that case the blue rectangle would have looked luminous.

2. For all three rectangles to appear as holes, three different illuminations have to be assumed, none of which has a common component characteristic as they are illuminating nothing more in the visual field than one single rectangle.

An assumption about illumination as a common component and an assumed minimum principle easily explains Metelli's demonstration.

Walter Gerbino has objected to my explanation of Metelli's demonstration and suggested a better example. It is a nice experiment by Koffka (1935, p. 256ff), where he placed two identical gray disks in adjacent rooms with a hole in the wall between the rooms. In one of the rooms the illumination was colored (e.g., red) and in the other it was white. Because of the hole in the wall the observer could see both disks simultaneously. Now the disk in the red illumination appeared gray (color constancy) but the disk in the white light appeared greenish. When looking through a reduction screen with two peepholes making only the two disks visible, the disk in red light of course appeared red and the one in white light appeared gray (i.e., no constancy). I am very grateful to Gerbino for bringing this experiment to my attention.

3.4. Simulated Illumination

In the Bergström (1980, 1982) experiments, cast shadows simulated attached shadows, thus inducing three-dimensional form. But illumination and shadow can also be simulated by modulation of reflectance. That is a technique applied in photos, drawings, and paintings. A painter can paint a motif as if in a certain illumination (e.g., by giving it a common hue component, and the shadows its complementary hue). Very interesting examples of simulated illumination are the constancy experiments by Arend (chap. 4, Fig. 4.6) and his co-workers (e.g., Arend & Goldstein, 1989), where he simulated different illuminations of two Mondrians on a computer screen by giving them different luminance levels. As long as the common component criterion is fulfilled, we perceive illumination no matter whether the manipulated physical variable is luminance, reflectance, or illumination.

3.5. Conclusion

We have found much evidence supporting our main assumption that variation of illumination is distinguished from variations of reflectance thanks to its characteristics of being a common component. The support comes from the classical studies of constancy, from our own experiments, and from Metelli's elegant demonstrations to the ECVP conference in Italy.

4. COLOR CONSTANCY AND DEPTH PERCEPTION

So far we have been discussing color constancy as a question of veridical color perception. When simulating attached shadows, as in the experiment mentioned earlier, we touched on the problem of distinguishing a shadow from a darker surface color. This distinction is essential for veridical color and depth perception.

4.1. Color Constancy and Perceived Solid Shape

The connection between color constancy and depth perception was demonstrated by Mach (1959) using a visiting card folded at a right angle and illuminated on one half and shaded on the other. When seen as flat the card appeared to have different colors on the two halves, but when seen as folded the two halves looked the same color but differently lit (i.e., the color and form percepts were veridical). This dependence between color constancy and depth perception has inspired many experiments. Some of the studies show little or no effect of perceived depth on perceived color; others show considerable effects. The latter have sometimes been considered to support a cognitive cascade type of theory, where one percept depends on another one. Most often color perception has been considered to depend on depth perception rather than the other way around. (cf. Epstein, 1961; Gilchrist, 1977, 1979, 1980; Hochberg & Beck, 1954; Rock, 1975, 1977).

4.1.1. Gilchrist's Studies.
A series of ingenious experiments in this area has been reported by Gilchrist (1977, 1980). He raised the question (1980): "When does perceived lightness depend on perceived spatial arrangement?"

In one class of experiments reported by Gilchrist, no effect of perceived depth on perceived lightness was observed. The experiments are variations of Mach's card experiment. The arrangements are shown in Fig. 6.9. The illumination ratio between the upper and lower half of the card was always 30:1. Gilchrist used four stimulus displays. In three of them the upper and lower half of the card (the upper and lower target) had the same reflectance (white, gray, or black), whereas in the fourth display they had different reflectances (upper black, lower white) but equal luminance.

One group of subjects observed the targets monocularly (which means that the two targets appeared coplanar), and the other group observed them binocularly (which means that the two halves of the card appeared perpendicular to each other). The observers made matchings of the two targets to Munsell samples.

The results showed no difference between the monocular and the binocular conditions for any of the four displays. Luminance differences between the targets were always perceived as lightness differences (i.e., there was no color constancy). The perceived depth had no effect on perceived lightness.

In the second class of experiments the perceived depth seems to have a dramatic effect on perceived lightness. The targets were coplanar (with one exception), but they were presented in front of a background that was

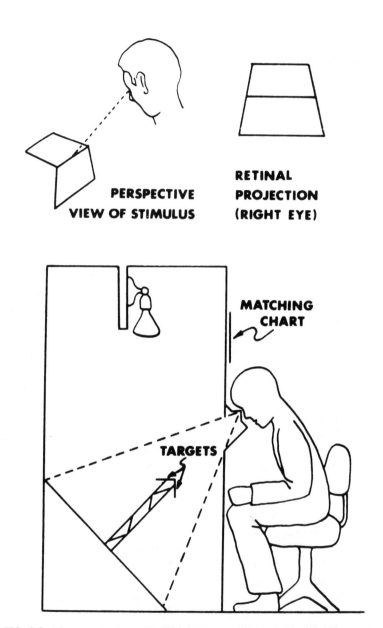

FIG. 6.9. Arrangements used in Gilchrist's perpendicular planes experiments (from Gilchrist, 1980).

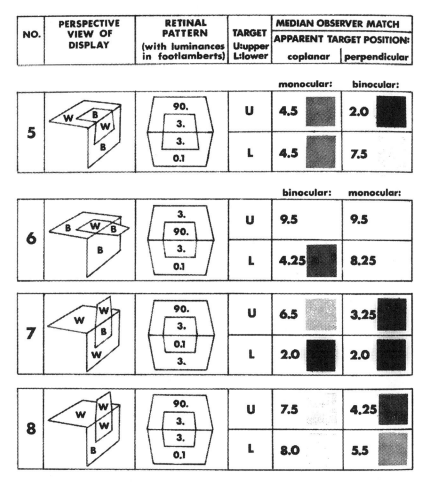

FIG. 6.10. Stimulus displays and matching data for perpendicular planes experiments (from Gilchrist, 1980).

folded into two perpendicular halves, one horizontal, the other vertical. The illumination ratio between the upper (horizontal) and the lower (vertical) half was 30.1. The four displays are shown in Fig. 6.10 and so are the main results. (Observe that for Displays 6 to 8 the monocular condition will make the targets appear perpendicular, whereas the binocular condition will make them appear coplanar, i.e., horizontal in 6 and vertical in 7 and 8).

The results are very clear-cut. In the binocular condition reflectance differences between upper and lower target are manifested in the Munsell matchings, whereas in the monocular condition the lower black target of Display 6 appears white and the upper white target of 7 appears black, a.s.o.

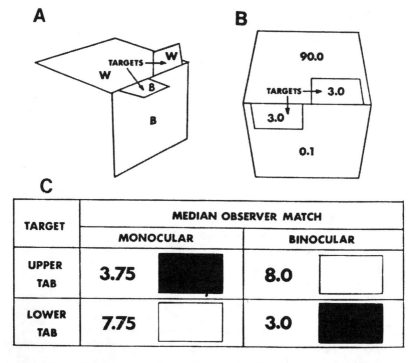

FIG. 6.11. (A) Perspective view of the stimulus display used in the critical test showing color (B = black; W = white) of each part. (B) Monocular retinal pattern showing luminances in foot-Lamberts. (C) Average Munsell matches for monocular and binocular viewing conditions (from Gilchrist, 1980).

Another dramatic example is given by the experiment illustrated in Fig. 6.11, where a change of perceived spatial arrangement is followed by a shift from black to white of one and the same target.

When comparing the two classes of experiments, it is essential to consider the difference in luminance ratios between the stimulus displays of the two classes. *In the first class* of experiments the luminance ratio within any single display is restricted to 30:1, which is the maximum ratio possible using reflected light and one common illumination (if fluorescent painting and specular reflectances are avoided). *In the second class* luminance ratios as high as 900:1 are used, which means that two separate illuminations are needed—either one for the upper target and its background and another for the lower target and its background, or one source of light illuminating one background and the other background being a source of light itself.

The conditions giving the dramatic shifts in perceived lightness all have two highly different illuminations. The target is perceived to be in one illumination when viewed monocularly and in the other when viewed binocularly. This means that highly different common components are extracted from one and the same target leaving highly different relative components left. This, of course, will give different perceived color of the target according to our model.

4.1.2. Bergström's Experiments.

My co-workers and I (Bergström, 1980, 1982, 1984; Bergström, Gustafsson, & Jakobsson, 1993; Bergström et al., 1980, 1984; Gustafsson & Bergström, 1987; Gustafsson et al., 1987a, 1987b) reported experiments where spatial and spatiotemporal variations of illumination make flat surfaces appear as three-dimensional objects, as walls forming a corner, as swinging doors, and so forth. And these phenomena are working against the traditional depth cues, against Gogel's (1965) equidistance tendency, and with free binocular observation.

It was shown earlier (Bergström, 1980, 1982) that putting the superimposed grating out of focus, thus giving gradual rather than stepwise variation of illumination, gives very compelling depth effects. The Mondrian can appear as a series of pipes, one above the other, and moving the grating a little may cause the pipes to appear to rotate in an induced motion. With luminance gradients (penumbras) the identification of shadows is unambiguous and the perceived pipes are always convex.

The grating in sharp focus makes the percept ambiguous, however, as we saw in the experiments reported earlier. Illumination steps can be mistaken for reflectance steps, and the surface can appear either flat or pleated. Thus, the percepts may be manipulated by manipulating the stimulus target (e.g., by varying its structure, the spatial frequency of the grating and of the Mondrian, the number of cycles presented a.s.o). A number of such experiments have been performed by my co-workers and me, and they have been or will be reported elsewhere (Bergström et al., 1984; Gustafsson & Bergström 1987; Gustafsson et al., 1987a, 1987b).

4.1.3. The Direction of Illumination.

In the analysis of the luminance gradient (Fig. 6.1) and of the Land and McCann black-and-white Mondrian (Fig. 6.3 and 6.4), the extracted common components of illumination were gradients informing about the direction of illumination. In the simulated attached shadow experiments discussed earlier, there are no such gradients, but a perceived attached shadow indicates a directed light. With the grating horizontally oriented, the illumination could be from above or from below as illustrated in Fig. 6.12.

When the grating is vertical, the illumination could equally well be from the left or from the right.

Now, in one of our experiments (Bergström et al., 1984, exp. 2.) the observers were asked to tell from what direction the perceived three-dimensional object appeared to be illuminated. Only one grating contrast was used—the one earlier found to be optimal for three- dimensional percepts.

The results show that with the grating vertically oriented there was a

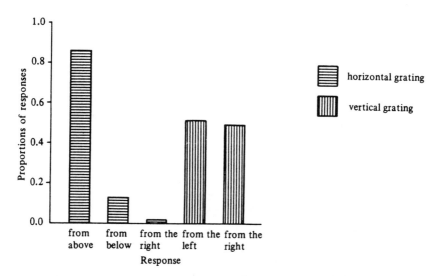

FIG. 6.12. Proportions of responses "illuminated from above," "from below," "from the right," and "from the left" for horizontal and vertical orientations of the grating (from Bergström et al., 1984, Experiment 2; reprinted from *Perception*, 1984, *13*, 129–140, Pion Limited, London).

perfect 50/50 split of answers "from the left" and "from the right." With the grating horizontally oriented, however, there was a very clear dominance of "from above" answers over "from below" answers (frequencies around .85 and .12, respectively). This preference for "from above" responses is in line with earlier observations concerning aerial photos, where concavities turn into convexities if the image is turned upside down (Hess, 1950, 1961; Oppel, 1856; Rittenhouse, 1786; Schröder, 1858; von Fieandt, 1938; von Fieandt and Moustgaard, 1977). These observations seem to imply the perceptual rule that in the absence of (unambiguous) proximal information about the direction of light, illumination is assumed to be from overhead. An elegant study by Herschberger (1970) seemed to show this assumption to be innate in chickens. A more recent study by Howard, Bergström, and Ohmi (1990) showed that the term *from overhead* is more adequate than *from above*, because adult observers have a strong tendency to judge the illumination as coming from overhead even when the head is horizontally oriented or even upside down.

4.2. Color Constancy and Perceived Separation in Depth

In the "square-wave illumination" experiments discussed earlier, the subjects were instructed to report three-dimensionality only when perceiving the target as one coherent three-dimensional object. It was observed in pilot studies, however, that at high contrasts of the grating there was a separation in depth. The target could appear like an illuminated picture or even a room seen through and behind a dark grating. This separation in depth is not included in the three-dimensional responses in the diagram of Fig. 6.8. The decrease of 3-D responses at higher contrast can essentially be explained as an effect of this instruction.

In later experiments our observers were instructed to use three categories when responding to the same type of stimuli as in the experiment discussed before. The responses were *flat, pleated,* and *separated in depth*. The results show "flat" responses to have their maximum at low grating

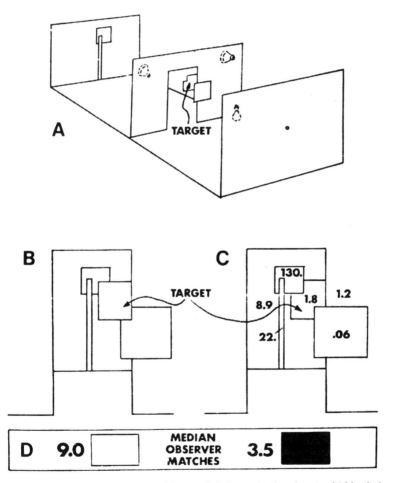

FIG. 6.13. (A) Perspective view of the parallel planes display showing hidden light bulbs. The display (as seen through the pinhole) in which the target appeared to be located either (B) in the near plane or (C) in the far plane, with luminances shown in foot-Lamberts. (D) The average match from a Munsell chart for the two displays (from Gilchrist, 1980).

contrast, "pleated" responses to have their maximum at medium contrast, and "separated" responses finally show a maximum at the highest contrasts (cf. Gustafsson & Bergström, 1987). Separation of colors in depth and also its dependence on contrast conditions is known from the literature (e.g., Bergström & Derefeldt, 1975; Fuchs, 1923; Katz, 1935; Metelli, 1970). The separation of color in depth raises the question of the effect of illumination on perceived depth as distinguished from perceived solid shape, which was discussed previously.

As a matter of fact, some of the Gilchrist experiments discussed earlier could qualify in this section as well. Let us consider still another of his experiments (Gilchrist, 1980), where he used a display consisting of parallel planes observed through a peephole and illuminated by hidden sources of light. The arrangement is illustrated in Fig. 6.13.

The target, a piece of white cardboard, could appear to be located either in the near plane or in the far plane, depending on manipulation of interpositioned objects. The far wall was illuminated by 71 times the intensity of that of the near wall. As can be seen from the data in Fig. 6.13, the same white target appeared almost black when perceived to be in the far position but white when in the near position.

4.3. Conclusions About Color and Depth

A number of studies dealing with the connection between color constancy and perceived depth have been reviewed. Gilchrist (1980) stated the question: "When does perceived lightness depend on perceived spatial arrangement?" It is clear from this title that the study is based on what I have called a cognitive cascade type of theory, where one percept causes another one.

We have already commented on the difference between Gilchrist's two classes of experiments, those showing no effect of changes of perceived depth on perceived lightness and those showing dramatic effects. The difference is that the second class of experiments work with luminance ratios of such a magnitude that two different illuminations are perceived, one for each of the backgrounds. The target is perceived in one illumination when viewed monocularly and in the other when viewed binocularly. (In the experiment referred to in Fig. 6.13, this change is made by manipulating the interposition of objects instead of retinal disparity.) The answer to Gilchrist's question is that he asked the wrong question. Perceived lightness does not depend on perceived spatial arrangements any more than perceived spatial arrangements depend on perceived lightness. The perceived lightness corresponds to the residual relative luminance components left when extracting the common component of illumination. What Gilchrist is doing in his experiments, when successful, is to create a situation where the whole illumination frame of reference is shifted between viewing conditions. The effect is analogous to the Gelb effect, although the shift of frame of reference is initiated by other means.

Gogel and Tietz (1976) and Gogel and Sharkey (1989) showed that an attentional shift of frame of reference influences induced motion. By shifting attention from one moving frame of reference to another one, the physically vertical motion track of the target shifts between appearing inclined to the right and inclined to the left, the proximal stimulation being invariant. The difference between this demonstration and the Gelb and Gilchrist effects is mainly that the attentional shift may not be as compelling as Gelb's inserted white strip or Gilchrist's disparity.

The reason that results like Gilchrist's have been considered to support a cascade type of theory is, of course, that depth perception within these short distances can be so easily influenced. It appears almost as an independent variable. We have to remember, however, that it is viewing conditions that are directly manipulated, *not* depth perception. What would we conclude if we could find an equally convenient way to influence color constancy with variation of perceived depth as an effect? How could we know then what is first, the egg or the hen? The simulated attached shadow experiments reported by Bergström and his collaborators certainly are steps in that direction. The illumination of flat targets is modulated in space or in space and time to give very compelling and vivid percepts of three-dimensional objects and events. The constancy of color is maintained at the cost of a nonveridical depth perception.

Apparently, the spatial and spatiotemporal distribution of light (the luminance distribution) across the visual field informs us about a three-dimensional layout, and apparently illumination and shadow can be identified with no separate depth information present (or to be correct, with all other information about depth being contradictory).

Now, if both the traditional geometric projection of the outlines of objects on the retina and the projected luminance distribution across the visual field can tell us about depth, how do we know that one is primary and the other is secondary (mediated)? Percept/percept relationships certainly raise problems for any theory of perception, but the solution cannot be that whatever cue perceptionists started to study is primary— *not for that reason*, anyway.

Discussing the color/depth relationship, Gilchrist (1980) also made the point that "a decision about lightness is the same as a decision about depth," thus retreating from his earlier point of view.

The percept/percept relationships are, of course, highly interesting and nobody can deny their existence. The alternatives presented in our analyses of the luminance profiles across the Mondrians (cf. Fig. 6.5, 6.6, 6.7, preceding) are examples of using the logic of such relations. The components were not chosen by chance among the infinite number of mathematical possibilities; they were chosen according to implicit rules.

We assume that perception is based on very complicated higher order physical conditions (complicated in our mathematical language), most of which we do not even know how to describe in a relevant way, plus organizing principles, which we do not know well either, but principles making the perceptual result so amazingly simple and logical. To use percept/percept relationships may then be the only simple way of describing and predicting perception. Gogel's equations relating, for instance, perceived head movement, perceived distance, and perceived object motion to each other are very elegant examples, as is the size/distance invariance hypothesis. But these equations do not imply a cause and effect relationship. They only formalize the restrictions, the logical covariation that seems to exist between certain aspects of percepts in specific situations. The same type of "perceptual logic" was used in the choice between alternatives in the analysis of the luminance profiles, though intuitively. This logic represents assumed organizing principles in our theory (decoding principles in Johansson's terminology, assumptions in the terms of the transactionalists, etc.). Most theorists probably agree that organizing principles are necessary to explain perception (orthodox Gibsonians and Watsonians excluded); the disagreement is about their origin.

5. UNECOLOGICAL SOURCES OF LIGHT

So far we have tried to explain data on color constancy—our own as well as data from the literature—by means of our model for automatic analysis of reflected light into common and relative components. Some of the data have been examples of color constancy, others have been examples of lack of constancy. Both have been explained by our model, either as examples of identification of variation of illumination thanks to conditions revealing its common component characteristic, or as examples of conditions concealing that same characteristic. Land and McCann's Mondrian demonstrations, Gilchrist's class of unsuccessful experiments, and our own demonstrations of three-dimensional percepts due to a modulated illumination of flat surfaces are all examples of color constancy that can be explained by our model. Gelb's and Kardos' experiments, Gilchrist's successful experiments, and Katz's reduction screen conditions are all examples of lack of constancy that can be explained as depending on lack of the commonality necessary for the variation of illumination to be identified. These conditions giving no or very little color constancy in the laboratory are not very often seen in real life. But we all know that there are a number of artificial sources of light that usually influence the perceived color of illuminated objects. There is a considerable literature on color rendering.

Color constancy is very good under natural conditions of illumination (i.e., where the source of light is a heated body or a heated body plus a "natural" filter, such as a cloud, which has a continuous transmittance spectrum). But as soon as we start using lamps with discontinuous or multimodal power spectra, color constancy becomes poorer. Many of our artificial light sources are unecological in the sense that they have spectral characteristics that would never occur in a natural environment. Most fluorescent tubes have a continuous power spectrum combined with a few power "spikes" at certain wavelengths. This combination means a high intensity per watt but at the same time a reduced color constancy. Our visual system is not programmed for such discontinuities in the power spectrum of the illuminant, and the extraction of the common component does not consider the spikes. Certain surface colors then become unreasonably luminous because of the spikes, and they may turn into luminous colors rather than surface colors. The "clearness" of colors in the illumination of certain "three-component illuminants" can be explained as depending on these unecological power spikes, which the visual system does not consider when extracting the common component of illumination. The "extra" intensity of some colors because of the discontinuities of the power spectrum is considered a relative rather than a common component (i.e., the relations appear as surface color differences rather than as a characteristic of the illumination). The situation resembles Metelli's local blue light that appeared to be a surface color rather than a hole into a box with a local blue illumination.

Even Land's Mondrian demonstrations show the same type of effect. There is a tremendous degree of color constancy in these demonstrations considering the drastic changes of the illumination. But some of the colors of the Mondrian as a matter of fact lose their surface-color characteristic and appear luminous.

6. OUR MODEL SUMMARIZED

6.1. Postulates and Assumptions

The first and main postulate is that the visual system performs a perceptual analysis of reflected light into common and relative components.

6.1.1. *The Common Component.* The common component is assumed to correspond to illumination (ilco).

It should be stressed here that "common" does not mean "uniform." Already in my first paper on this matter (Bergström, 1977b) an example was given where the common component was a gradient informing about the direction of illumination. In real life it is usually much more complex (a changing topography).

The modulation of illumination always depends on the 3D layout of the environment including the position of the source of light (and on nothing else). Anything that modulates the illumination is a solid object, a fluid, or a gas in space, and the resulting modulation informs us about it. With artificial light the varying distance from the source of light also modulates the illumination of surfaces (cf. the achromatic Mondrian illustration in Fig. 6.3 and 6.4).

6.1.2. The Relative Components. The relative components correspond to the variation of the materials illuminated, their changes in space and/or time as to degree of reflectance, selectivity of reflectance, and so on.

6.1.3. Two Intermixed Modulations. The reflected light has a modulation in space and/or time that is the product (vector sum?) of the modulation of the illumination and that of the reflectance. Modulation of illumination corresponds to the 3D layout, whereas modulation of reflectance corresponds to changes of color. These are our two components.

The perceptual system uses the analysis into common and relative components to discriminate between the two intermixed modulations in space and/or time. And as in motion perception, we assume that the fact that one is a common component and one is not is what distinguishes the two. The visual system is good at picking up common features and making them the background or the frame of reference for other events in the field. This is well known from the many studies of Johansson motions and from many of the traditional gestalt demonstrations.

6.2. Edges and Gradients

Gradual changes in space and/or time are frequent characteristics of illumination, whereas sharp edges normally belong to reflectance. The penumbra is an example of the former, borders between colors of the latter. A functional difference between edges and gradients was assumed in my early paper (Bergström, 1977b, p. 183) to explain the Craik–O'Brien–Cornsweet effect. As the luminance edge was assumed to correspond to a reflectance border, its effect could not be canceled or compensated for by a luminance gradient. According to our model, the gradient would make the (blacker) ring look curved in depth rather than restore its degree of whiteness to equal that of the first ring.

The assumption of a functional difference between edges and gradients still holds true in general. Our architecture and other artifacts have given us an environment with many sharp illumination edges, however. And to reveal the common component characteristics of these sharp illumination borders certain conditions must be fulfilled, conditions that the Craik–O'Brien–Cornsweet demonstrations normally do not fulfill. (For experimental studies on different conditions that help reveal the common component, i.e., illumination, characteristics of sharp illumination borders; see Bergström, 1982, Gustafsson & Bergström, 1987, and Gustafsson et al., 1987a, 1987b). In a recent experiment Bergström et al. (1993) studied the effect of the number of reflectance edges in the display to reveal the illumination characteristics of the luminance step in an "O'Brien effect" illumination. The central disk appeared more distinctly convex or concave (rather than darker gray) the higher the number of real reflectance edges that were crossed by the O'Brien contour.

6.3. The Minimum Principle

A minimum principle applied in our model is that a minimum but geometrically sufficient number of perceived sources of light is assumed.

The minimum principle was applied to Metelli's demonstration (aforementioned). Because there is a limit to the contrast that is possible between two surface colors in a common illumination (about 30:1), contrasts higher than that limit result in the perception of two sources of light, either one of the fields becomes luminous or the two fields are perceived to be in different illuminations (cf. the explanation of Gilchrist's data; see earlier).

6.4. Why Call it a Vector Model?

Even though no formal vector analysis has been performed and even though most of our experimental work has been performed under conditions where an episcotister or slice model (e.g., Gerbino, Stultiens, Troost, and de Weert, 1990; chap. 5) would do, there are at least three reasons to keep to the vector model.

1. The model emanates from Johansson's vector model for motion perception. Remember, there was no vector model for motion perception as long as motion was considered an inference from perception of static stimuli.

2. Alternative models are based on studies of evenly illuminated homogeneous fields or episcotister conditions. My model is supposed to apply to reality, where illuminations and luminances really are modulated.

3. This means that my common component is usually a complex pattern rather than a uniform light. As soon as we start letting this vary also in time to give perceived motion in depth, a vector model will certainly be necessary. Whether it will be sufficient still has to be tested.

So far very little experimental work has been performed with simultaneous variation of illumination (luminance) in both space and time (Bergström, 1982, 1984). But the little that has been done has shown very clearly that, as in motion perception, the variation in time gives very compelling perceptions with little or no ambiguity.

7. CONCLUSIONS

Empirical support for the vector model has been given both from the literature (from Katz, 1911, to Metelli, 1983) and from experimental studies performed in my own laboratory. The model (i.e., the commonality assumption plus a minimum principle as to the number of light sources) has been applied to a number of color (lightness) constancy studies (both constancy and nonconstancy results), to Land's Mondrian demonstrations, to Gilchrist's (1980) two classes of experiments, one class giving a radical shift of perceived lightness and the other giving lightness constancy, to Metelli's (1983) demonstration of a hole in a screen appearing like a surface color, and also to the problem of color rendering. The problem of color edges versus illumination edges is also solved within the model (Gustafsson, 1987, Gustafsson et al., 1987a, 1987b, and the discussion earlier).

ACKNOWLEDGMENTS

Unfortunately, I could not attend the Trieste symposium on August 25, 1987. This absence may account for part of the lack of coherence between my contribution and those of my co-authors. On the very day of the symposium and during the following weeks, Professor F. Camerini and his staff at the Ospedale Maggiore in Trieste were fighting to save as much as possible of my heart tissue. Had it not it been for their competence and devotedness, this chapter would never have been completed. Needless to say, I am very grateful to them.

I also want to express my gratitude to professor Walter C. Gogel, University of California, Santa Barbara, and to professor Gunnar Johansson, University of Uppsala, Sweden, for uncountable intense and interesting discussions of essential parts of this chapter.

This study was initially supported financially by grants from the Swedish Council for Research in the Humanities and Social Sciences under contracts No. F 255–83, F 146–84, and F 59–85.

Glossary

The following list of definitions apply mostly to the achromatic domain. "Nontechnical" means less precise description in ordinary language.

achromatic vs. chromatic: In the case of surfaces, the term achromatic refers to colors along the scale of grays from black to white, whereas chromatic refers to colors that vary in hue, saturation, and lightness. In the case of lights, achromatic means neutral.

brightness: Perceived luminance. The apparent amount of light coming to the eye from a region of the field.

color of light: Hue, saturation, and brightness. See also *surface color*.

contrast: Use of the term *contrast* without modifiers has produced a great deal of confusion in the literature due to its several very different usages. In this book we have made an effort to minimize the use of the term contrast alone.

    (a) physical contrast: relative luminance; common measures are Weber contrast: $\Delta L/L_b$ and Michelson contrast: $(L_{max} - L_{min})/(L_{max} + L_{min})$;

    (b) apparent contrast; the perceived amount of a luminance transition;

    (c) a set of perceptual phenomena in which the color of a visual region is altered in a direction away from that of either an adjacent (simultaneous contrast) or a preceding (successive contrast) visual region;

    (d) a theoretical mechanism or process proposed to explain (c). Historically, this use of the term has been associated with neural mechanisms like lateral inhibition.

Further discussion of these confusions can be found on pages 14 and 15.

contrast-brightness: a term used in chapters 2 and 3 for the dimension from black to glaring describing the apparent intensity of light coming from a region of the visual field with an illuminated surround.

illuminance: (technical) The illuminance at a point of a surface is the quotient of the luminous flux incident on an infinitesimal element of the surface containing the point under consideration, by the area of that surface element (Wyszecki & Stiles, 1967).

$$E = K_m \int_\lambda E_{e,\lambda} V(\lambda) \, d\lambda$$

where K_m is a units constant, $E_{e,\lambda}$ is spectral irradiance, and $V(\lambda)$ is the CIE photopic luminous efficiency function (Wyszecki & Stiles, 1967).

illumination: the nontechnical word for illuminance.

lightness: perceived reflectance. The dimension of perceived surface color that in the achromatic domain ranges from white to black. In the chromatic domain, lightness refers to the intensitive dimension.

lightness constancy: The stability of the perceived lightness of a surface despite changes in the proximal stimulus. These proximal changes are caused by factors such as changes in the level of illumination, changes in the background of a surface, and changes of orientation of the surface.

luminance: (technical) The luminance at a point of a surface and in a given direction is the quotient of the luminous intensity in the given direction of an infinitesimal element of the surface containing the point under consideration, by the orthogonally projected area of the surface element on a plane perpendicular to the given direction (Wyszecki & Stiles, 1967)

$$L = K_m \int_\lambda L_{e,\lambda} V(\lambda) \, d\lambda$$

where K_m is a units constant, $L_{e,\lambda}$ is spectral radiance, and $V(\lambda)$ is the CIE photopic luminous efficiency function (Wyszecki & Stiles, 1967).

(nontechnical) physical amount of light reaching a viewpoint from a given region of the optic array; sometimes referred to as intensity. Note that intensity has a technical definition as a description of point sources (Wyszecki & Stiles, 1967).

luminance edge vs. luminance gradient: in nontechnical usage, edge typically refers to an abrupt spatial transition from one luminance to another. Gradient refers to a gradual spatial change of luminance. Gradient also has a specific mathematical definition: the directional rate of change over space or time. See chapter 4.

reflectance: (technical) The ratio of the reflected radiant flux (or power) to incident radiant flux (or power) (Wyszecki & Stiles, 1967).

Spectral Reflectance:

$$\rho(\lambda) = \frac{P_\lambda}{P_{0\lambda}}$$

where π_λ is the reflected spectral power and $\pi_{0\lambda}$ is the incident spectral power.

Luminous Reflectance:

$$\rho = \frac{\int_\lambda \rho(\lambda) V(\lambda) P_{0\lambda} \, d\lambda}{\int_\lambda V(\lambda) P_{0\lambda} \, d\lambda}$$

where $\pi_{0\lambda}$ is the incident spectral power, $\rho(\lambda)$ is spectral reflectance, and $V(\lambda)$ is the CIE luminous efficiency function.

(nontechnical) percentage of light a surface reflects; physical blackness or whiteness of a surface. *Reflectance*, unless otherwise specified, means luminous reflectance.

surface color: Surface hue, surface saturation, and lightness. There are no terms in widespread usage to distinguish between colors of surfaces and chromatic colors of lights, analogous to the lightness/brightness distinction. It is therefore necessary to use terms such as *surface hue, surface saturation*, and *lightness*. See also *color of light*.

References

Adelson, E. H., & Pentland, A. P. (1990). *The perception of shading and reflectance* (Vision and Modeling Tech. Rep. No. 140). Cambridge, MA: MIT Media Laboratory.

Adelson, E. H., Pentland, A., & Kuo, J. (1989). The extraction of shading and reflectance. *Investigative Opthalmology and Visual Science, 30*(3), 262.

Aguilar, M., & Stiles, W. S. (1954). Saturation of the rod mechanism of the retina at high levels of stimulation. *Optica Acta, 1*, 59–65.

Alpern, M. (1964). Relation between brightness and color contrast. *Journal of the Optical Society of America, 54*, 1491–1492.

Arend, L. E. (1973). Spatial differential and integral operations in human vision: Implications of stabilized retinal image fading. *Psychological Review, 80*, 374–395.

Arend, L. E. (1976). Temporal determinants of the form of the spatial contrast threshold MTF. *Vision Research, 16*, 1035–1042.

Arend, L. E. (1985). Spatial gradient illusions and inconsistent integrals. *Investigative Opthalmology and Visual Science, 26*, 280.

Arend, L. (1991). Apparent contrast and surface color in complex scenes. In B. E. Rogowitz, M. H. Brill, & J. P. Allebach (Eds.), *Human vision, visual processing, and digital display II, Proceedings of SPIE 1453* (pp. 412–421). San Jose, CA: SPIE.

Arend, L. (1993a). How much does illuminant color affect unattributed colors? *Journal of the Optical Society of America A, 10*(10), 2134–2147.

Arend, L. E. (1993b). Mesopic brightness, lightness, and brightness-contrast. *Perception and Psychophysics, 54*(10), 469–476.

Arend, L. E. (In preparation). *Both lightness and brightness are affected by background reflectance*.

Arend, L. E., Buehler, J. N., & Lockhead, G. R. (1971). Difference information in brightness perception. *Perception and Psychophysics, 9*, 367–370.

Arend, L., & Goldstein, R. (1987a). Lightness models, gradient illusions, and curl. *Perception and Psychophysics, 42,* 65–80.

Arend, L., & Goldstein, R. (1987b). Simultaneous constancy, lightness and brightness. *Journal of the Optical Society of America A, 4,* 2281–2285.

Arend, L., & Goldstein, R. (1989). Lightness and brightness in unevenly illuminated scenes. In M. Pietikainen (Ed.), *Proceedings of the 6th Scandinavian Conference on Image Analysis* (pp. 499–506). Oulu, Finland.

Arend, L., & Goldstein, R. (1990). Lightness and brightness over spatial illumination gradients. *Journal of the Optical Society of America A, 7*(10), 1929–1936.

Arend, L., & Reeves, A. (1986). Simultaneous color constancy. *Journal of the Optical Society of America A, 3,* 1743–1751.

Arend, L., Reeves, A., Schirillo, J., & Goldstein, R. (1991). Simultaneous color constancy: Patterns with diverse Munsell values. *Journal of the Optical Society of America A, 8*(4), 661–672.

Arend, L., & Spehar, B. (1993a). Lightness, brightness and brightness contrast. I. Illumination variation. *Perception and Psychophysics, 54*(10), 446–456.

Arend, L., & Spehar, B. (1993b). Lightness, brightness and brightness contrast. II. Reflectance variation. *Perception and Psychophysics, 54*(10), 457–468.

Arend, L., & Timberlake, G. (1986). What is psychophysically perfect image stabilization? Do perfectly stabilized images always disappear? *Journal of the Optical Society of America A, 3,* 235–241.

Barlow, R. B., & Verillo, R. T. (1976). Brightness sensation in a ganzfeld. *Vision Research, 16,* 1291–1297.

Barrow, H. G., & Tenenbaum, J. (1978). Recovering intrinsic scene characteristics from images. In A. R. Hanson & E. M. Riseman (Eds.), *Computer vision systems* (pp. 3–26). Orlando, FL: Academic Press.

Barrow, H. G., & Tenenbaum, J. (1986). Computational approaches to vision. In K. R. Boff, L. Kaufman, & J. P. Thomas (Eds.), *Handbook of perception and human performance* (pp. 38.1–38.70). New York: Wiley.

Bartleson, C. J. (1977a). *Factors affecting color appearance and measurement by psychophysical methods.* Unpublished doctoral thesis, The City University, London.

Bartleson, C. J. (1977b). A review of chromatic adaptation. In F. W. Billmeyer & G. Wyszecki (Eds.), *AIC Proceedings, Color 77* (pp. 63–96). Bristol, England: Adam Hilger.

Battersby, W. S., & Wagman, I. W. (1964). Light adaptation kinetics: The influence of spatial factors *Science, 143,* 1029–1031.

Beck, J. (1972). *Surface color perception.* Ithaca, NY: Cornell University Press.

Beck, J., Prazdny, K., & Ivry, R. (1984). The perception of transparency with achromatic colors. *Perception and Psychophysics, 35,* 407–422.

Bergström, S. S. (1977a). *Colour.* Stockholm: Swedish Broadcasting Company.

Bergström, S. S. (1977b). Common and relative components of reflected light as information about the illumination, colour, and three-dimensional form of objects. *Scandanavian Journal of Psychology, 18,* 180–186.

Bergström, S. S. (1980). *Illumination, colour, and 3-D form* (Department of Applied Psychology Reports (DAPS) No. 3). University of Umeå, Sweden.

Bergström, S. S. (1982). Illumination, color, and three-dimensional form. In J. Beck (Ed.), *Organization and representation in perception* (pp. 365–378) Hillsdale, NJ: Lawrence Erlbaum Associates.

Bergström, S. S. (1984). Modulations of illumination giving a flat surface apparent three-dimensional shape and motion in depth. *Perception, 13,* 1.
Bergström, S. S., & Derefeldt, G. (1975). Effects of surround/test field luminance ratio on induced colour. *Scandanavian Journal of Psychology, 16,* 311–318.
Bergström, S. S., Gustafsson, K. A., & Jakobsson, T. (1991). *Distinctness of perceived three-dimensional form induced by modulated illumination: II. Effects of the distinctness of perceived depth in the O'Brien phenomenon* (No. 35). Department of Applied Psychology, University of Umeå, Sweden.
Bergström, S.S., Gustafsson, K.-A., Jakobsson, T. (1993). Distinctness of perceived three-dimensional form induced by modulated illumination: Effects of certain display and modulation conditions. *Perception & Psychophysics, 53* (6), 648– 657.
Bergström, S. S., Gustafsson, K.-A., & Putaansuu, J. (1980, September). Distribution of reflected light analyzed into components providing information about 3-D shape and illumination. *Third European Conference on Visual Perception,* Brighton, England.
Bergström, S. S., Gustafsson, K.-A., & Putaansuu, J. (1984). Information about three-dimensional shape and direction of illumination in a square-wave grating. *Perception, 13,* 129–140.
Bezold, W. v. (1876). *The theory of color and its relation to art and art-industry* (S. R. Koehler, Trans.). Boston: Sprang (Original work published 1874.)
Björklund, R. A., & Magnussen, S. (1979). Decrement version of the Broca-Sulzer effect and its spatial analogue. *Vision Research, 19,* 155–157.
Blake, A. (1985a). Boundary conditions for lightness computation in Mondrian world. *Computer Vision, Graphics, and Image Processing, 32,* 314–327.
Blake, A. (1985b). On lightness computation in the Mondrian world. In D. Ottoson & S. Zeki (Eds.), *Central and peripheral mechanisms of colour vision* (pp. 45–59). New York: MacMillan.
Blake, A., & Bülthoff, H. (1990). Does the brain know the physics of specular reflection? *Nature, 343*(6254), 165–168.
Bohm, D. (1965). *The special theory of relativity.* New York: Benjamin.
Börjesson, E., & von Hofsten, C. (1972). Spatial determinants of depth perception in two-dot motion patterns. *Perception and Psychophysics, 11*(4), 263–268.
Börjesson, E., & von Hofsten, C. (1973). Visual perception of motion in depth: Application of a vector model to three-dot motion patterns. *Perception and Psychophysics, 13,* 169–179.
Börjesson, E., & von Hofsten, C. (1975). A vector model for perceived object rotation and translation in space. *Psychological Research, 5,* 209–230.
Brainard, D. H., & Wandell, B. A. (1986). Analysis of the retinex theory of color vision. *Journal of the Optical Society of America A, 3,* 1651–1661.
Brenner, E., Cornelissen F., & Nuboer, W. (1987) *Some spatial aspects of simultaneous colour contrast.* Pre-conference abstract for Seeing Contour and Colour conference, Manchester, UK.
Brill, M. H. (1978). Letters to the editor: A device performing illuminant-invariant assessment of chromatic relations. *Journal of Theoretical Biology, 71,* 473–478.
Brill, M. H. (1979). Computer simulation of object-color recognizers. *Journal of Theoretical Biology, 78,* 305–308.
Brill, M. H. (1984). Physical and informational constraints on the perception of transparency and translucency. *CVGIP, 28,* 356–362.

Brindley, G. S. (1960). *Physiology of the retina and visual pathway*. London: Edward Arnold.
Brookes, A., & Stevens, K. A. (1989). The analogy between stereo depth and brightness. *Perception, 18*, 601–614.
Brown, J. L., & Mueller, C. G. (1965). Brightness discrimination and brightness contrast. In C. H. Graham (Ed.), *Vision and visual perception* (pp. 208–250) New York: Wiley.
Buchsbaum, G. (1980). A spatial processor model for object color perception. *Journal of the Franklin Institute, 310*, 1–26.
Bülthoff, H. H., & Mallot, H. A. (1987). Interaction of different modules in depth perception. *Proceedings of the IEEE 1st International Conference on Computer Vision*, 295–304.
Bülthoff, H. H., & Mallot, H. A. (1988). Integration of depth modules: Stereo and shading. *Journal of the Optical Society of America A, 5*(10), 1749–1758.
Burgh, P., & Grindley, G. C. (1962). Size of test patch and simultaneous contrast. *Quarterly Journal of Experimental Psychology, 14*, 89–93.
Burkhardt, D. A., Gottesman, J., Kersten, D., & Legge, G. E. (1984). Symmetry and constancy in the perception of negative and positive luminance contrast. *Journal of the Optical Society of America A, 1*, 309–316.
Burnham, R. W. (1953). Bezold's color-mixture effect. *American Journal of Psychology, 66*, 377–385.
Burnham, R. W., Evans, R. M., & Newhall, S. M. (1957). Prediction of color appearance with different adaptation illuminants. *Journal of the Optical Society of America, 47*, 35–42.
Campbell, F. W., & Howell, E. R. (1972). Monocular alternation: A method for the investigation of pattern vision. *Journal of Physiology, 225*, 19–21P.
Campbell, F. W., Johnstone, J. R., & Ross, J. (1981). An explanation for the visibility of low frequency gratings. *Vision Research, 21*, 723–730.
Carter, R. (1993). Gray scale and achromatic color difference. *Journal of the Optical Society of America A,10*, 1380–1391.
Chen, B., MacLeod, D. I. A., & Stockman, A. (1987). Improvement of human vision under bright light: Grain or gain? *Journal of Physiology, 394*, 41–66.
Chubb, C., Sperling, G., & Solomon, J. A. (1989). Texture interactions determine perceived contrast. *Proceedings of the National Academy of Sciences USA, 86*, 9631–9635.
CIE. (1970). *International lighting vocabulary* (3rd ed.). Paris: CIE Central Bureau.
Cohn, T. E., & Lasley, D. J. (1986). Visual sensitivity. *Annual Review of Psychology, 37*, 495–521.
Cornsweet, T. N. (1970). *Visual perception*. New York: Academic Press.
Cornsweet, T. N., & Teller, D. Y. (1965). Relation of increment thresholds to brightness and luminance. *Journal of the Optical Society of America, 55*, 1303–1308.
Craik, K. J. W. (1940). The effect of adaptation on subjective brightness. *Proceedings of the Royal Society B, 128*, 232–247.
Cramer, T. (1923). Über die Beziehung des Zwischenmediums zu den Transformations- und Kontrasterscheinungen [The effect of intervening media on transformation- and contrast-appearances]. *Zeitschrift für Sinnesphysiologie, 54*, 215–242.
Cutting, J. E. (1978). Generation of synthetic male and female walkers through manipulation of a biomechanical invariant. *Perception, 7*, 393–405.

Dannemiller, J. L. (1989). Computational approaches to color constancy: Adaptive and ontogenetic considerations. *Psychological Review, 96,* 255–266.
Davidson, M., & Whiteside, J. A. (1971). Human brightness perception near sharp contours. *Journal of the Optical Society of America, 61,* 530–536.
Dennett, D. (1988). Quining qualia. In A. J. Marcel & E. Bisiach (Eds.), *Consciousness in contemporary science* (pp. 42–77). Oxford: Clarendon Press.
Dennett, D. (1991). *Consciousness explained.* London: Allen Lane.
DeValois, R. L., Webster, M. A., DeValois, K. K., & Lingelbach, B. (1986). Temporal properties of brightness and color induction *Vision Research, 26,* 887–897.
Diamond, A. L. (1962). Brightness of a field as a function of its area. *Journal of the Optical Society of America, 52,* 700–706.
du Buf, J. M. H. (1987). *Spatial characteristics of brightness and apparent-contrast perception.* Unpublished doctoral thesis, Institute for Perception Research, Eindhoven, The Netherlands.
D'Zmura, M., & Lennie, P. (1986). Mechanisms of color constancy. *Journal of the Optical Society of America A, 3,* 1662–1672.
Eastman Kodak. (1986). *Kodak color films and papers for professionals.* Rochester, NY: Author.
Eisner, A., & MacLeod, D. I. A. (1980). Blue cones do not contribute to luminance. *Journal of the Optical Society of America, 70,* 121–123.
Epstein, W. (1961). Phenomenal orientation and perceived achromatic color. *Journal of Psychology, 52,* 51–53.
Epstein, W. (1977). Historical introduction to the constancies. In W. Epstein (Ed.), *Stability and constancy in visual perception: Mechanisms and processes* (pp. 1–22). New York: Wiley.
Evans, R. M. (1943). Visual processes and color photography. *Journal of the Optical Society of America, 33*(11), 579–614.
Evans, R. M. (1948). *An introduction to color.* New York: Wiley.
Evans, R. M. (1974). *The perception of color.* New York: Wiley.
Fairchild, M. D., & Lennie, P. (1992). Chromatic adaptation to natural and incandescent illuminants. *Vision Research, 32*(11), 2077–2085.
Festinger, L., Coren, S., & Rivers, G. (1970). The effect of attention on brightness contrast and assimilation. *American Journal of Psychology, 83,* 189–207.
Fiorentini, A., Baumgartner, G., Magnussen, S., Schiller, P. H., & Thomas, J. P. (1990). The perception of brightness and darkness: Relations to neuronal receptive fields. In L. Spillmann & J. Werner (Eds.), *Visual perception: The neurophysiological foundations* (pp. 129–161). San Diego: Academic Press.
Flock, H. R. (1974). Stimulus structure in lightness and brightness experiments. In R. MacLeod & H. Pick (Eds.), *Perception: Essays in honor of James J. Gibson* (pp. 185–208). Ithaca, NY: Cornell University Press.
Flock, H. R., & Noguchi, K. (1970). An experimental test of Jameson and Hurvich's theory of brightness contrast. *Perception and Psychophysics, 8,* 129–136.
Forsyth, D. A. (1988) *A new algorithm for colour constancy.* Unpublished doctoral thesis, Oxford University, Oxford, UK.
Forsyth, D. A., & Zisserman, A. (1990). Shape from shading in the light of mutual illumination. *Image and Vision Computing, 8*(1), 42–49.
Freeman, R. B. (1967). Contrast interpretation of brightness constancy. *Psychological Bulletin, 67,* 165–187.

Friden, T. P. (1973). Whiteness constancy: Inference or insensitivity? *Perception and Psychophysics, 1*, 81–89.

Fry, G. A. (1948). Mechanisms subserving simultaneous brightness contrast. *American Journal of Optometry, 25*, 162–178.

Fuchs, W. (1923). Experimentelle Untersuchungen uber das simultane Hintereinandersehen auf derselben Sehrichtung. , *Abt. I*(91), 145–235.

Gelb, A. (1929). Die "Farbenkonstanz" der Sehdinge. In W. A. von Bethe (Ed.), *Handbuch norm. und pathol. Psychologie* (pp. 594–678).

Georgeson, M. A., & Sullivan, G. D. (1975). Contrast constancy: Deblurring in human vision by spatial frequency channels. *Journal of Physiology, 252*, 627–656.

Gerbino, W. (1985). An ecological model of achromatic transparency. In *Third International Conference on Event Perception and Action*, University of Uppsala, Sweden.

Gerbino, W. (1988). Models of achromatic transparency: A theoretical analysis. *Gestalt Theory, 10*, 5–20.

Gerbino, W., Stultiens, C., Troost, J., & de Weert, C. (1990). Transparent layer constancy. *Journal of Experimental Psychology: Human Perception and Performance, 16*(1), 3–20.

Gerrits, H. J. M., & Vendrik, A. J. H. (1970). Simultaneous contrast, filling-in process and information processing in man's visual system. *Experimental Brain Research, 11*, 411–430.

Gibson, J. J. (1966). *The senses considered as perceptual systems*. Boston: Houghton Mifflin.

Gibson, J. J. (1975). Three kinds of distance that can be seen, or how Bishop Berkeley went wrong. In G. B. Flores d'Arcais (Ed.), *Studies in perception: Festschrift for Fabio Metelli* (pp. 83–87). Milano: Martello-Giunti.

Gibson, J. J. (1979). *The ecological approach to visual perception*. Boston: Houghton Mifflin.

Gilchrist, A. L. (1977). Perceived lightness depends on perceived spatial arrangement. *Science, 195*, 185–187.

Gilchrist, A. L. (1979). The perception of surface blacks and whites. *Scientific American, 240*, 112–123.

Gilchrist, A. L. (1980). When does perceived lightness depend on perceived spatial arrangements. *Perception and Psychophysics, 28*, 527–538.

Gilchrist, A. L. (1988). Lightness contrast and failures of constancy: A common explanation. *Perception and Psychophysics, 43*(5), 415–424.

Gilchrist, A. L. & Bonato, F. (In press). Anchoring of lightness values in center/surround displays. *Journal of Experimental Psychology: Human Perception and Performance*.

Gilchrist, A. L., Delman, S., & Jacobsen, A. (1983). The classification and integration of edges as critical to the perception of reflectance and illumination. *Perception and Psychophysics, 33*, 425–436.

Gilchrist, A. L., & Jacobsen, A. (1983). Lightness constancy through a veiling luminance. *Journal of Experimental Psychology: Human Perception and Performance, 9*, 936–944.

Gilchrist, A. L., & Jacobsen, A. (1984). Perception of lightness and illumination in a world of one reflectance. *Perception, 13*, 5–19.

Gilchrist, A. L., & Jacobsen, A. (1988). Qualitative relationships are decisive. *Perception and Psychophysics, 45*(1), 92–94.

Glad, A., Magnussen, S., & Engvik, H. (1976). Temporal brightness and darkness enhancement: Further evidence for asymmetry. *Scandanavian Journal of Psychology, 17*, 234–237.

Gogel, W. C. (1965). Equidistance tendency and its consequences. *Psychological Bulletin, 64*, 153–163.

Gogel, W. C., & Sharkey, T. J. (1989). Measuring attention using induced motion. *Perception, 18*, 303–320.

Gogel, W. C., & Tietz, J. D. (1976). Adjacency and attention as determiners of perceived motion. *Vision Research, 16*, 839–845.

Graham, M. E., & Rogers, B. J. (1982). Simultaneous and successive contrast effects in the perception of depth from motion parallax and stereoscopic information. *Perception, 11*, 247–262.

Graham, N. V. S. (1989). *Visual pattern analyzers.* New York: Oxford University Press.

Grossberg, S., & Todorovic, D. (1988). Neural dynamics of 1-d and 2-d brightness perception: A unified model of classical and recent phenomena. *Perception and Psychophysics, 43*, 241–277.

Gustafsson, K.-A. (1987). *Perceived three-dimensional form induced by modulated illumination* (21). University of Umeå, Sweden, Department of Applied Psychology (DAPS).

Gustafsson, K., & Bergström, S. S. (1987). *Conditions influencing the 3-D form induced by modulated illumination* (18). University of Umeå, Sweden, Department of Applied Psychology (DAPS).

Gustafsson, K., Bergström, S. S., & Jakobsson, T. (1987a). *Conditions influencing the 3-D form induced by modulated illumination. II(a): Illumination borders and reflectance borders* (19). University of Umeå, Sweden, Department of Applied Psychology (DAPS).

Gustafsson, K., Bergström, S. S., & Jakobsson, T. (1987b). *Conditions influencing the 3-D form induced by modulated illumination. II(b): Paired comparison studies of colour and structure* (20). University of Umeå, Sweden, Department of Applied Psychology (DAPS).

Haimson, B. R. (1974). The response criterion, the stimulus configuration, and the relationship between brightness contrast and brightness constancy. *Perception and Psychophysics, 16*(2), 347–354.

Hallett, P., Jepson, A., & Gershon, G. (1988). Colour constancy at the level of the cones. *Investigative Opthalmology and Visual Science, 29*(163).

Hamada, J. (1984). A multi-stage model for border contrast. *Biological Cybernetics, 51*, 65–70.

Hasegawa, T. (1977). Simultaneous color contrast: Deviations of the induced colors from directions of complementary colors. In F. W. Billmeyer & G. Wyszecki (Eds.), *Color 77.* Bristol, England: Adam Hilger.

Hatfield, G., & Epstein, W. (1985). The status of the minimum principle in the theoretical analysis of visual perception. *Psychological Bulletin, 97*, 155–186.

Hayhoe, M., & Wenderoth, P. (1991). Adaptation mechanisms in color and brightness. In A. Valberg & B. Lee (Eds.), *From pigments to perception* (pp. 353–367). New York: Plenum.

Hayhoe, M., Wenderoth, P., Lynch, E., & Brainard, D. (1991). Adaptation mechanisms in color appearance. *Investigative Opthalmology and Visual Science, 32*(4).

Hayhoe, M. M., Benimoff, N. I., & Hood, D. C. (1987). The time-course of multiplicative and subtractive adaptation process. *Vision Research, 27*(11), 1981–1996.

Hayhoe, M. M., Levin, M. E., & Koshel, R. J. (1992). Subtractive processes in light adaptation. *Vision Research, 32,* 323–333.

Hayhoe, M. M., & Whittle, P. (1992). The time course of adaptational influences on the brightness of flashes. *Investigative Opthalmology and Visual Science, 33,* 4.

Heeger. D. J. (1992). Normalization of cell responses in cat striate cortex. *Visual Neuroscience, 9,* 181–197.

Heggelund, P. (1974). Achromatic color vision—II: Measurement of simultaneous achromatic contrast within a bidimensional system. *Vision Research, 14,* 1081–1088.

Heggelund, P. (1991). On achromatic colors. In A. Valberg & B. Lee (Eds.), *From pigments to perception* (pp. 313–323). New York: Plenum.

Heider, G. M. (1933). New studies in transparency, form, and color. *Psychologische Forschung, 17,* 13–56.

Heinemann, E. G. (1955). Simultaneous brightness induction as a function of inducing-and test-field luminances. *Journal of Experimental Psychology, 50,* 89–96.

Heinemann, E. G. (1961). The relation of apparent brightness to the threshold for differences in luminance. *Journal of Experimental Psychology, 61,* 389–399.

Heinemann, E. G. (1972). Simultaneous brightness induction. In J. D. Hurvich & L. M. Hurvich (Eds.), *Handbook of sensory physiology.* Berlin: Springer.

Helmholtz, H. (1962). *Helmholtz's treatise on physiological optics.* (J. P. C. Southall, Trans.; 3rd ed.). New York: Dover. (Original work published 1866)

Helson, H. (1938). Fundamental problems in color vision. I. The principle governing changes in hue, saturation and lightness of nonselective samples in chromatic illumination. *Journal of Experimental Psychology, 23,* 439–436.

Helson, H. (1943). Some factors and implications of color constancy. *Journal of the Optical Society of America, 33*(10), 555–567.

Helson, H. (1963). Studies of anomalous contrast and assimilation. *Journal of the Optical Society of America, 53,* 179–84.

Helson, H., & Jeffers, V. B. (1940). Fundamental problems in color vision. II. Hue, lightness, and saturation of selective samples in chromatic illumination. *Journal of Experimental Psychology, 26,* 1–27.

Helson, H., Judd, D. B., & Warren, M. H. (1952). Object-color changes from daylight to incandescent filament illumination. *Illumination Engineering, 47,* 221.

Helson, H., & Michels, W. C. (1948). The effect of chromatic adaptation on achromaticity. *Journal of the Optical Society of America, 38,* 1025–1032.

Henneman, R. H. (1935). A photometric study of the perception of object color. *Archives of Psychology* (No. 179), 5–89.

Hering, E. (1890). Beitrag zur Lehre vom Simultankontrast. [A contribution to the laws of simultaneous contrast]. *Zeitschrift für Psychologie, 1,* 18–28.

Hering, E. (1964). *Outlines of a theory of the light sense.* (L. M. Hurvich & D. Jameson, Trans.). Cambridge, MA: Harvard University Press. (Original work published 1874)

Herschberger, W. (1970). Attached-shadow orientation perceived as depth by chickens reared in an environment illuminated from below. *Journal of Comparative Physiology and Psychology, 73,* 407–411.

Hess, C. & Pretori, H. (1970). Quantitative investigation of the lawfulness of simultaneous brightness contrast (H. Flock & J. H. Tenny, Trans.). *Perceptual and Motor Skills 31,* 947–969. (Original work published 1984).

Hess, E. H. (1950). Development of the chick's response to light and shade cues of depth. *Journal of Comparative Physiology and Psychology, 43,* 112–122.

Hess, E. H. (1961). Shadows and depth perception. *Scientific American, 204,* 139–148.

Hochberg, J. E., & Beck, J. (1954). Apparent spatial arrangement and perceived brightness. *Journal of Experimental Psychology, 47,* 263–266.

Holmgren, S. (1974). *On searching for Mach type phenomena in the visual perception of spatial velocity distributions, 51.* University of Uppsala, Department of Psychology.

Horn, B. K. P. (1974). Determining lightness from an image. *Computer Graphics and Image Processing, 3,* 277–299.

Horn, B. K. P. (1977). Understanding Image Intensities. *Artificial Intelligence, 8,* 201–231.

Howard, I., Bergström, S. S., & Ohmi, M. (1990). Shape from shading in different frames of reference. *Perception, 19,* 523–530.

Hsia, Y. (1943). Whiteness constancy as a function of difference in illumination. *Archives of Psychology* (284).

Hunt, R. W. G. (1950). The effects of daylight and tungsten-light adaptation on color perception. *Journal of the Optical Society of America, 40,* 362–371.

Hurvich, L. M., & Jameson, D. (1966). *The perception of brightness and darkness.* Boston: Allyn & Bacon.

Jacobsen, A., & Gilchrist, A. (1988a). Hess and Pretori revisited: Resolution of some old contradictions. *Perception and Psychophysics, 43,* 7–14.

Jacobsen, A., & Gilchrist, A. (1988b). The ratio principle holds over a million-to-one range of illumination. *Perception and Psychophysics, 43,* 1–6.

Jameson, D., & Hurvich, L. M. (1961). Complexities of perceived brightness. *Science, 133,* 174–179.

Jameson, D., & Hurvich, L. M. (1964). Theory of brightness and color contrast in human vision. *Vision Research, 4,* 135–154.

Johansson, G. (1950). *Configurations in event perception.* Uppsala, Sweden: Almqvist & Wiksell.

Johansson, G. (1964). Perception of motion and changing form. *Scandanavian Journal of Psychology, 5,* 181–208.

Johansson, G. (1975). Visual motion perception. *Scientific American,* 76–88.

Johansson, G. (1976). Spatio-temporal differentiation and integration in visual motion perception. *Psychological Research, 38,* 379–393.

Johansson, G. (1977). Spatial constancy and motion in visual perception. In W. Epstein (Ed.), *Stability and constancy in visual perception. Mechanisms and processes* (pp. 375–420). New York: Wiley.

Johansson, G. (1982). Visual space perception through motion. In A. H. Wertheim, W. A. Wagenaar, & E. H. W. Leibowitz (Eds.), *Tutorials on motion perception* (pp. 19–39). New York: Plenum.

Judd, D. B. (1940). Hue saturation and lightness of surface colors with chromatic illumination. *Journal of the Optical Society of America, 30,* 2–32.

Judd, D. B. (1960). Appraisal of Land's work on two-primary color projections. *Journal of the Optical Society of America, 50,* 254–268.

Judd, D. B., & Wyszecki, G. (1975). *Color in business, science, and industry.* New York: Wiley.

Jung, R. (1973). Visual perception and neurophysiology: Central processing of visual information. In R. Jung (Ed.), *Handbook of sensory physiology.* (pp. 1–152). Berlin: Springer.

Kanizsa, G. (1979). *Organization in vision.* New York: Praeger.

Kanizsa, G., & Luccio, R. (1986). Die Doppeldeutigkeiten der Prägnanz [The ambiguity of prägnanz]. *Gestalt Theory, 8,* 99–135.

Kardos, L. (1929). Die "Konstanz" phänomenaler Dingmomente [The constancy of perceived things]. *Beitrage Problemgeschichte Psychologie, (Bühler Festschrift),* 1–77.

Kardos, L. (1934). Ding und Schatten [Object and shadow]. *Zeitschrift Psychologie, Erg. bd 23.*

Katz, D. (1911). Die Erscheinungsweisen der Farben und ihre Beeinflussung durch die individuelle Erfahrung [The appearance of colors and its dependence on the individual experience]. *Zeitschrift für Psychologie, Erg. bd 7.*

Katz, D. (1935). *The world of colour.* (R. B. MacLeod & C. W. Fox, Trans.). London: Kegan Paul, Trench, Trubner & Co.

Kelly, D. H. (1979a). Motion and vision. I. Stabilized images of stationary gratings. *Journal of the Optical Society of America, 69,* 1266–1274.

Kelly, D. H. (1979b). Motion and vision. II. Stabilized spatio-temporal threshold surface. *Journal of the Optical Society of America, 69,* 1340–1349.

Kingdom, F., & Moulden, B. (1991). A model for contrast discrimination with incremental and decremental test patches. *Vision Research, 31*(4), 851–858.

Kinney, J. A. S. (1962). Factors affecting induced color. *Vision Research, 2,* 503–525.

Kinney, J. A. S. (1965). Effect of exposure time on induced color. *Journal of the Optical Society of America, 55,* 731–736.

Klinker, G. J. (1988) *A physical approach to color image understanding.* Unpublished doctoral thesis, Carnegie-Mellon University, Pittsburgh, PA.

Koenderink, J. J., van de Grind, W. A., & Bouman, M. A. (1971). Foveal information processing at photopic luminances. *Kybernetik, 8,* 128–144.

Koffka, K. (1931). Some remarks on the theory of colour constancy. *Psychologische Forschung, 16,* 329–354.

Koffka, K. (1935). *Principles of gestalt psychology.* New York: Harcourt, Brace, & World.

Köhler, W. (1915). Optische Untersuchungen am Schimpansen und am Haushuhn. In *Abhandlungen der Koniglich Preussischen Akademie der Wissenschaften.*

Köhler, W. (1971). On unnoticed sensations and errors of judgment. In M. Henle (Ed.), *The selected papers of Wolfgang Köhler.* New York: Liveright.

Kozaki, A. (1973). Perception of lightness and brightness of achromatic surface color and impression of illumination. *Japanese Psychological Research, 15,* 194–203.

Krauskopf, J. (1963). Effect of retinal image stabilization on the appearance of heterochromatic targets. *Journal of the Optical Society of America, 53,* 741–744.

Laming, D. R. J. (1986). *Sensory analysis*. New York: Academic Press.
Land, E. H. (1974). The Retinex theory of colour vision. *Proceedings of the Royal Institute of Great Britain, 47*, 23–58.
Land, E. H. (1977). The retinex theory of color vision. *Scientific American, 237*, 108–128.
Land, E. H. (1983). Recent advances in retinex theory and some implications for cortical computations: Color vision and the natural image. *Proceedings of the National Academy of Sciences USA, Physics, 80*, 5163–5169.
Land, E. H., & McCann, J. J. (1971). Lightness and retinex theory. *Journal of the Optical Society of America, 61*, 1–11.
Leeuwenberg, E. L. J. (1978). Quantification of certain visual pattern properties: Salience, transparency, similarity. In E. L. J. Leeuwenberg & H. F. J. M. Buffart (Eds.), *Formal theories of perception* (pp. 277–298). New York: Wiley.
Leeuwenberg, E. L. J., & Buffart, H. F. J. M. (1983). An outline of coding theory: A summary of related experiments. In H. Geissler, H. F. J. M. Buffart, E. L. J. Leeuwenberg, & V. Sarris (Eds.), *Modern issues in perception* (pp. 25–47). Amsterdam: North-Holland.
Le Grand, Y. (1957). *Light, colour and vision*. London: Chapman & Hall.
Leshowitz, B., Taub, H. B., & Raab, D. H. (1968). Visual detection of signals in the presence of continuous and pulsed backgrounds. *Perception and Psychophysics, 4*, 207–213.
Lie, I. (1969). Psychophysical invariants of achromatic colour vision. I. The multidimensionality of achromatic colour experience. *Scandinavian Journal of Psychology, 10*, 167–175.
Lie, I. (1977). Perception of illumination. *Scandinavian Journal of Psychology, 18*, 251–255.
MacAdam, D. L. (1935). Maximum visual efficiencies of colored materials. *Journal of the Optical Society of America, 25*, 361–367.
Mach, E. (1865). On the effect of the spatial distribution of the light stimulus on the retina. In F. Ratliff (Ed.), *Mach bands: Quantitative studies on neural networks in the retina* (pp. 253–271). San Francisco: Holden-Day.
Mach, E. (1959). *Die Analyse der Empfindungen [The analysis of sensations]*. New York: Dover. (Original work published 1922)
MacLeod, R. B. (1932). An experimental investigation of brightness constancy. *Archives of Psychology, 135*, 5–102.
MacLeod, R. B. (1940). Brightness constancy in unrecognized shadows. *Journal of Experimental Psychology, 27*(1), 1–22.
MacLeod, R. B., & Pick, H. L. (Eds.). (1974). *Perception. Essays in honor of James J. Gibson*. Ithaca, NY: Cornell University Press.
Maffei, L., & Fiorentini, A. (1972). Retinogeniculate convergence and analysis of contrast. *Journal of Neurophysiology, 35*, 65–72.
Magnussen, S., & Glad, A. (1975a). Effects of steady surround illumination on the brightness and darkness enhancement of flickering lights. *Vision Research, 15*, 1413–1416.
Magnussen, S., & Glad, A. (1975b). Temporal frequency characteristics of spatial interaction in human vision. *Experimental Brain Research, 23*, 519–528.

Maloney, L. T. (1985). Computational approaches to color constancy. In *Stanford Applied Psychology Laboratory Technical Report* Stanford, CA: Stanford University.
Maloney, L. T., & Wandell, B. A. (1986). Color constancy: A method for recovering surface spectral reflectance. *Journal of the Optical Society of America A, 3*, 29–33.
Marks, L. E. (1974). Blue-sensitive cones can mediate brightness. *Vision Research, 14*, 1493–1494.
Marmolin, H. (1973a). Visually perceived motion in depth resulting from proximal changes I. *Perception and Psychophysics, 14*, 133–142.
Marmolin, H. (1973b). Visually perceived motion in depth resulting from proximal changes II. *Perception and Psychophysics, 14*, 143–148.
Marmolin, H. (1977). Three- and two-dimensional motion after-effects. *Scandinavian Journal of Psychology, 18*, 192–202.
Marr, D. (1978). Representing visual information. In A. R. Hanson & E. M. Riseman (Eds.), *Computer vision systems* (pp. 61–80). Orlando, FL: Academic Press.
Marr, D. (1982). *Vision*. San Francisco: Freeman.
Masin, S. C. (1984). An experimental comparison of three- versus four-surface phenomenal transparency. *Perception and Psychophysics, 35*, 325–332.
Mausfeld, R., & Niederee, R. (1993). An inquiry into relational concepts of colour based on incremental principles of colour coding for minimal relational stimuli. *Perception, 22*, 427–462.
Maximov, V. V. (1984). *Transformation of colors by illuminant change*. Moscow: Nayka.
McCann, J. J., McKee, S. P., & Taylor, T. H. (1976). Quantitative studies in retinex theory. *Vision Research, 16*, 445–458.
Merleau-Ponty, M. (1945). *Phenomenology of perception*. London: Kegan, Paul.
Metelli, F. (1970). An algebraic development of the theory of perceptual transparency. *Ergonomics, 13*, 59–66.
Metelli, F. (1975a). The perception of transparency. In G. B. Flores d'Arcais (Ed.), *Studies in perception: Festschrift for Fabio Metelli* (pp. 445–487). Milano: Martello-Giunti.
Metelli, F. (1975b). Shadows without penumbra. In Ertel, Kemmler, & Stadler (Eds.), *Gestaltheorie in der modernen Psychologie* [Gestalt theory in modern psychology] (pp. 200–209). Darmstadt: Steinkopff.
Metelli, F. (1976). What does "more transparent" mean? A paradox. In M. Henle (Ed.), *Vision and artifact* (pp. 19–24). New York: Springer.
Metelli, F. (1983). Three perceptual illusions: Their meaning for perceptual theory. *Perception, 12*, A1–A40.
Metelli, F. (1985). Stimulation and perception of transparency. *Psychological Research, 7*, 185–202.
Metelli, F., Da Pos, O., & Cavedon, A. (1985). Balanced and unbalanced, complete and partial transparency. *Perception and Psychophysics, 38*, 354–366.
Metzger, W. (1934a). Betrachtungen uber phänomenale Identitet. *Psychologische Forschung, 19*, 1–60.
Metzger, W. (1934b). Tiefenerscheinungen in optischen Bewegungsfeldern. *Psychologische Forschung, 20*, 195–206.
Mingolla, E. (1983). *Perception of shape and illuminant direction from shading*. Unpublished doctoral thesis, University of Connecticut, Storrs, CT.

Mingolla, E., & Todd, J. T. (1986). Perception of solid shape from shading. *Biological Cybernetics, 53*, 137–151.

Mullen, K. T. (1985). The contrast sensitivity of human colour vision to red-green and blue-yellow chromatic gratings. *Journal of Physiology, 359*, 381–400.

Munsell, A. E. O., Sloan, L. L., & Godlove, I. H. (1933). Neutral value scales. I. Munsell Neutral Value Scale. *Journal of the Optical Society of America, 23*, 394–411.

Nachmias, J., & Steinman, R. M. (1965). Brightness and discriminability of light flashes. *Vision Research, 5*, 545–557.

Nakayama, K., Shimojo, S., & Ramachandran, V. S. (1990). Transparency: Relation to depth, subjective contours, luminance and neon color spreading. *Perception, 19*, 497–513.

Neisser, U., & Becklen, R. (1975). Selective looking: Attending to visually-specified events. *Cognitive Psychology, 5*, 480–494.

Neri, D. (1989) *The chromatic Cornsweet effect: Relationships between spatial and chromatic variables.* Unpublished doctoral thesis, University of Connecticut, Storrs, CT.

Nikolaev, P. P. (1988a). Algorithms for color discrimination of objects according to the reactions of logarithmic receptors. *Biofizika, 33*, 517–521.

Nikolaev, P. P. (1988b). Monocular color discrimination of nonplanar objects under various illumination conditions. *Biofizika, 33*, 140–144.

Noguchi, K., & Masuda, N. (1971). Brightness changes in a complex field with changing illumination: A re-examination of Jameson and Hurvich's study of brightness constancy. *Japanese Psychological Research, 13*, 60–69.

Ogawa, T., Kozaki, T., Takano, T., and Okayama, K. (1966) Effect of area on apparent brightness. Report of the Psychological Laboratory, Keio University, Japan.

Oppel, J. J. (1856). Uber ein Anaglyptoskop [On the stereoscope]. *Annalen der Physik und Chemie, 175*, 466–469.

Oppenheim, A. V., Schafer, R. W., & Stockham, T. G., Jr. (1968). Nonlinear filtering of multiplied and convolved signals. *Proceedings of the IEEE, 56*, 1264–1291.

Parrish, M., & Smith, K. (1967). Simultaneous brightness contrast as a function of perceptual set. *Psychonomic Science, 7*, 155–156.

Pentland, A. P. (1989). Local shading analysis. In B. K. P. Horn & M. J. Brooks (Eds.), *Shape from shading* (pp. 443–487). Cambridge, MA: MIT Press.

Piantanida, T., & Gilchrist, A. (In preparation). Direct evidence for edge integration in lightness perception.

Pokorny, J., Shevell, S. K., & Smith, V. C. (1991). Colour appearance and colour constancy. In P. Gouras (Ed.), *Vision and visual dysfunction* (pp. 43–61). London: Macmillan.

Poulton, E. C. (1989). *Bias in quantifying judgments.* Hillsdale, NJ: Lawrence Erlbaum Associates.

Ratliff, F. (1965). *Mach bands: Quantitative studies on neural networks in the retina.* San Francisco: Holden-Day.

Remondino, C. (1975). A model for the prediction of the conditions for phenomenal transparency. In G. B. Flores d'Arcais (Ed.), *Studies in perception: Festschrift for Fabio Metelli* (pp. 111–138). Milano: Martello-Giunti.

Restle, F. (1979). Coding theory of the perception of motion configurations. *Psychological Review, 86*(1), 1–24.

Rittenhouse, D. (1786). Explanation of an optical deception. *Transactions of the American Philosophical Society, 2,* 37–42.
Rivers, W. H. R. (1900). Vision. In Schäfer (Ed.), *Textbook of physiology* (Vol. 2). Edinburgh: Young J. Pentland.
Robson, J. G. (1966). Spatial and temporal contrast sensitivity functions of the visual system. *Journal of the Optical Society of America, 56,* 1141–1142.
Rock, I. (1975). *An introduction to perception.* New York: Macmillan.
Rock, I. (1977). In defense of unconscious inference. In W. Epstein (Ed.), *Stability and constancy in visual perception: Mechanisms and processes* (pp. 321–374). New York: Wiley.
Rock, I. (1983). *The logic of perception.* Cambridge, MA: MIT Press.
Ronchi, V. (1983). *Storia della luce.* Bari: Laterza.
Rozhkova, G. I., Nickolayev, P. P., & Shchardrin, V. E. (1982). Perception of stabilized retinal stimuli in dichoptic viewing conditions. *Vision Research, 22,* 293–302.
Runesson, S., & Frykholm, G. (1983). Kinematic specification of dynamics as an informational basis for person-and action perception: Expectation, gender, recognition, and deceptive intention. *Journal of Experimental Psychology: Human Perception and Performance, 112,* 585–615.
Sällström, P. (1973). *Some remarks concerning the physical aspect of human colour vision* (No. 73-09). Stockholm: University of Stockholm Institute of Physics.
Schröder, H. (1858). Über eine optische Inversion bei Betrachtung verkehrter, durch optische Vorrichtung entworfener physischer Bilder [On an optical inversion produced by viewing reversed stereoscopic images]. *Annalen der Physik und Chemie, 181,* 298–311.
Schubert, J., & Gilchrist, A. L. (1992). Relative luminance is not derived from absolute luminance. *Investigative Opthalmology and Visual Science, 33* (4), 1258.
Semmelroth, C. C. (1970). Prediction of lightness and brightness on different backgrounds. *Journal of the Optical Society of America, 60,* 1685.
Semmelroth, C. C. (1971). Adjustment of the Munsell value and W* scale to uniform lightness steps for various background reflectances. *Applied Optics, 10,* 14.
Sewall, L., & Wooten, B. R. (1991). Stimulus determinants of achromatic constancy. *Journal of the Optical Society of America A, 8,* 1794–1809.
Shafer, S. A. (1985). Using color to separate reflection components. *Color Research and Application, 10,* 210–218.
Shapley, R. (1986). The importance of contrast for the activity of single neurons, the VEP and perception. *Vision Research, 26*(1), 45–61.
Shapley, R., & Enroth-Cugell, C. (1984). Visual adaptation and retinal gain controls. *Progress in Retinal Research, 3,* 263–343.
Shapley, R., & Reid, R. C. (1985). Contrast and assimilation in the perception of brightness. *Proceedings of the National Academy of Sciences USA, 82,* 5983–5986.
Sharpe, L. T., Fach, C., Nordby, K., & Stockman, A. (1989). The incremental threshold of the rod visual system and Weber's law. *Science, 244,* 354–356.
Sharpe, L. T., Whittle, P., & Nordby, K. (1993). Spatial integration and sensitivity changes in the human rod visual system. *Journal of Physiology,*
Shevell, S. K. (1980). Unambiguous evidence for the additive effect in chromatic adaptation. *Vision Research, 20,* 637–639.

Shevell, S. K. (1986). On neural signals that mediate brightness. *Vision Research, 26*(8), 1195–1208.
Shevell, S. K., Holliday, I., & Whittle, P. (1992). Two separate neural mechanisms of brightness induction. *Vision Research, 32*, 2331–2340.
Sobagaki, H., Yamanaka, T., Takahama, K., & Nayatani, Y. (1974). Chromatic-adaptation study by subjective-estimation method. *Journal of the Optical Society of America, 64*, 743–749.
Spencer, D. E. (1943). Adaptation in color space. *Journal of the Optical Society of America, 33*, 10–17.
Spillmann, L., & Werner, J. S. (1990). *Visual perception: The neurophysiological foundations.* New York: Academic Press.
Stevens, S. (1961). To honor Fechner and repeal his law. *Science, 133*, 80–86.
Stockham, T. G. (1972). Image processing in the context of a visual model. *Proceedings of the IEEE, 828*–842.
Stromeyer, C. F. I., Cole, G. R., & Kronauer, R. E. (1985). Second-site adaptation in the red-green opponent pathway. *Vision Research, 25*, 219–237.
Takasaki, H. (1966). Lightness change of grays induced by change in reflectance of gray background. *Journal of the Optical Society of America, 56*, 504.
Thijssen, J. M., & Vendrik, A. J. H. (1971). Differential luminance sensitivity of the human visual system. *Perception and Psychophysics, 10*, 58–64.
Thouless, R. H. (1931). Phenomenal regression to the real object. *British Journal of Psychology, 21*, 339–359.
Todd, J. (1985). The perception of structure from motion: Is projective correspondence of moving elements a necessary condition? *Journal of Experimental Psychology: Human Perception and Performance, 11*, 689–710.
Todd, J. T., & Mingolla, E. (1983). Perception of surface curvature and direction of illumination from patterns of shading. *Journal of Experimental Psychology: Human Perception and Performance, 9*(4), 583–595.
Tuijl, H. F. J. M., & Leeuwenberg, E. L. J. (1979). Neon color spreading and structural information measures. *Perception and Psychophysics, 25*, 269–284.
Ullman, S., & Schechtman, G. (1982). Adaptation and gain normalization. *Proceedings of the Royal Society of London B, 216*, 299–313.
Valberg, A. (1974). Color induction: Dependence on luminance, purity and dominant or complementary wavelength of inducing stimuli. *Journal of the Optical Society of America, 64*, 1531–1540.
Valberg, A., & Lange-Malecki, B. (1990). "Colour constancy" in mondrian patterns: A partial cancellation of physical chromaticity shifts by simultaneous contrast. *Vision Research, 30*(3), 371–380.
Valberg, A., & Lee, B. (1991). *From pigments to perception.* New York: Plenum.
van der Helm, P. A., & Leeuwenberg, E. L. J. (1988). *Accessibility, a criterion for the choice of visual coding rules* (NICI Internal Report No. 01). NICI.
van der Wildt, G. J., Keemink, C. J., & van den Brink, G. (1976). Gradient detection and contrast transfer by the human eye. *Vision Research, 16*, 1047–1054.
von Fieandt, K. (1938). *Über Sehen von Tiefengebilden bei wechselnder Beleuchtungsrichtung* [The visual perception of depth under changing direction of illumination]. (Report of the Psychology Institute). University of Helsinki.
von Fieandt, K. (1949). Das phänomenologische Problem von Licht und Schatten

[The phenomenalogical problem of light and shadow]. *Acta Psychologica, 6,* 337–357.
von Fieandt, K., & Moustgaard, I. K. (1977). *The perceptual world.* London: Academic Press.
von Kries, J. (1904). Die Gesichtsempfindungen [The sensations of vision]. *Handbuch d. Physiol. d. Menschen, 3,* 211.
Vos, J. J., & Walraven, P. L. (1970). On the derivation of the foveal receptor primaries. *Vision Research, 11,* 795–818.
Wallach, H. (1948). Brightness constancy and the nature of achromatic colors. *Journal of Experimental Psychology, 38,* 310–324.
Wallach, H. (1963). The perception of neutral colors. *Scientific American, 208,* 107–116.
Wallach, H. (1976). *On perception.* New York: Quadrangle/The New York Times Book Co.
Wallach, H., & Galloway, A. (1946). The constancy of colored objects in colored illumination. *Journal of Experimental Psychology, 36,* 119–126.
Wallach, H., & O'Connell, D. N. (1953). The kinetic depth effect. *Journal of Experimental Psychology, 45*(4), 205–217.
Walls, G. L. (1954). The filling-in process. *American Journal of Optometry, 31,* 329–341.
Walls, G. L. (1960). Land! Land! *Psychological Bulletin, 57,* 29–48.
Walraven, J. (1976). Discounting the background—The missing link in the explanation of chromatic induction. *Vision Research, 16,* 289–295.
Walraven, J. (1981). Perceived colour under conditions of chromatic adaptation: Evidence for gain control by pi mechanisms. *Vision Research, 21,* 611–620.
Walraven, J., Benzschawel, T., Rogowitz, B., & Lucassen, M. P. (1991). Testing the contrast explanation of colour constancy. In A. Valberg & B. Lee (Eds.), *From pigments to perception* (pp. 369–377). New York: Plenum.
Walraven, J., Enroth-Cugell, C., Hood, D., MacLeod, D., & Schnapf, J. (1990). The control of visual sensitivity: Receptoral and postreceptoral processes. In L. Spillmann & J. Werner (Eds.), *The neurophysiological foundations of visual perception* (pp. 53–101). San Diego, CA: Academic Press.
Ware, C., & Cowan, W. B. (1983). The chromatic Cornsweet effect. *Vision Research, 23,* 1075–1077.
Waygood, M. (1969). The visibility of rate of change of luminance in the presence or absence of a boundary. *Optica Acta, 16,* 61–64.
Wertheimer, M. (1923). Untersuchungen zur Lehre von der Gestalt [Investigations in Gestalt-theory]. *Psychologische Forschung, 4,* 301–350.
Wertheimer, M. (1955). Gestalt theory. In W. D. Ellis (Ed.), *A source book of gestalt psychology* (pp. 1–11). New York: The Humanities Press.
Whittle, P. (1963). *Binocular rivalry.* Unpublished doctoral thesis, Cambridge University, Cambridge, UK.
Whittle, P. (1965). Binocular rivalry and the contrast at contours. *Quarterly Journal of Experimental Psychology, 17,* 217–226.
Whittle, P. (1973). The brightness of coloured flashes on backgrounds of various colours and luminances. *Vision Research, 13,* 621–638.
Whittle, P. (1974a). Intensity discrimination between flashes which do not differ in

brightness: Some new measurements on the "Blue" cones. *Vision Research, 14,* 599–601.

Whittle, P. (1974b). Luminance discrimination of separated flashes: The effect of background luminance and the shapes of T.V.I. curves. *Vision Research, 14,* 713–719.

Whittle, P. (1986). Increments and decrements: Luminance discrimination. *Vision Research, 26,* 1677–1691.

Whittle, P. (1991). Sensory and perceptual processes in seeing brightness and lightness. In A. Valberg & B. Lee (Eds.), *From pigments to perception* (pp. 293–304). New York: Plenum.

Whittle, P. (1992). Brightness, discriminability and the 'Crispening Effect'. *Vision Research, 32*(8), 1493–1507.

Whittle, P. (In preparation). The effect of background reflectance on lightness. *Vision Research*.

Whittle, P., & Arend, L. (1991). Homochromatic colour induction. *Perception, 20,* 99.

Whittle, P., & Challands, P. D. C. (1969). The effect of background luminance on the brightness of flashes. *Vision Research, 9,* 1095–1110.

Whittle, P., & Swanston, M. T. (1974). Luminance discrimination of separated flashes: The effect of background luminance and the shapes of T.V.I. curves. *Vision Research, 14,* 713–719.

Williams, L. G. (1966). *A study of image enhancement using eye movement measurement* (Unpublished Honeywell Rep. No. 12540-FR1). Minneapolis, MN: Honeywell.

Wilson, M. H., & Brocklebank, R. W. (1960). Two-colour projection phenomena. *Journal of Photographic Science, 8,* 141–149.

Witkin, A. P., & Tenenbaum, J. M. (1983). On the role of structure in vision. In J. Beck, B. Hope, & A. Rosenfeld (Eds.), *Human and machine vision* (pp. 481–544). Orlando, FL: Academic Press.

Wittgenstein, L. (1953). *Remarks on colour.* Oxford, England: Anscombe.

Woodworth, R. S. (Ed.). (1938). *Experimental psychology.* New York: Holt.

Woodworth, R. S., & Schlosberg, H. (1955). *Experimental psychology* (3rd ed.). London: Methuen.

Worthey, J. A. (1985). Limitations of color constancy. *Journal of the Optical Society of America A, 2,* 1014–1026.

Worthey, J. A. (1989). Geometry and amplitude of veiling reflections. *Journal of the Illumination Engineering Society, 18,* 49–62.

Wyszecki, G. (1986). Color appearance. In K. R. Boff, L. Kaufman, & J. P. Thomas (Eds.), *Handbook of perception and human performance* (pp. 1–57). New York: Wiley.

Wyszecki, G., & Stiles, W. S. (1967). *Color science.* New York: Wiley.

Yarbus, A. L. (1967). *Eye movements and vision.* New York: Plenum.

Young, T. (1807). *A course of lectures on natural philosophy and the mechanical arts.* London: J. Johnson, St Paul's Church Yard.

Zucker, S. W. (1987). The emerging paradigm of computational vision. *Annual Review of Computational Science, 2,* 69–89.

Author Index

A

Adelson, E. H., 29, 144
Aguilar, M., 48
Alpern, M., 94
Arend, L. E., 5, 6, 11, 16, 22, 25, 33, 95, 106, 113, 114, 115, 121, 122, 125, 129, 130, 134, 135, 136, 138, 141, 159, 160, 163, 164, 166, 167, 179, 183, 185, 186, 188, 189, 190, 191, 192, 193, 194, 198, 201, 202, 209, 210, 211, 212, 272

B

Barlow, R. B., 5
Barrow, H. G., 12, 13, 106, 143, 168, 169, 170
Bartleson, C. J., 155
Battersby, W. S., 105
Baumgartner, G., 36
Beck, J., 18, 37, 172, 179, 217, 218, 227, 231, 239, 243, 245, 246, 247, 250, 273
Becklen, R., 12
Benimoff, N. I., 98, 102, 104, 105, 107
Benzschawel, T., 94
Bergström, S. S., 20, 28, 33, 34, 93, 173, 192, 220, 257, 258, 259, 260, 261, 264, 266, 267, 269, 270, 272, 277, 278, 280, 283, 284, 285
Bezold, W. v., 65
Björklund, R. A., 52, 62, 91
Blake, A., 134, 159, 194, 199
Bohm, D., 171
Bonato, F., 15
Börjesson, E., 265

Bouman, M. A., 86
Brainard, D. H., 129, 202, 212
Brenner E., 21, 129
Brill, M. H., 204, 217, 227, 243, 246, 254
Brocklebank, R. W., 93
Brookes, A., 37
Brown, J. L., 38
Buchsbaum, G., 204
Buehler, J. N., 25, 136, 138, 192
Buffart, H. F. J. M., 221
Bülthoff, H., 199
Burgh, P., 132
Burkhardt, D. A., 70
Burnham, R. W., 133, 137, 138

C

Campbell, F. W., 92, 138, 192
Carter, R., 88
Cavedon, A., 218, 228, 230
Challands, P. D. C., 6, 21, 25, 26, 29, 37, 39, 43, 45, 56, 90, 116, 126, 132, 133, 138, 143, 173, 182, 188
Chen, B., 45, 53, 56, 57
Chubb, C., 100
Cohn, T. E., 76
Cole, G. R., 94
Coren, S., 137
Cornelissen, F., 21, 129
Cornsweet, T. N., 4, 14, 15, 16, 18, 19, 20, 32, 38, 72, 163
Cowan, W. B., 212
Craik, K. J. W., 50
Cramer, T., 127
Cutting, J. E., 265

309

D

Da Pos, O., 218, 228, 230
Dannemiller, J. L., 202
Davidson, M., 194
de Weert, C., 11, 199, 255
Delman, S., 2, 11, 16, 19, 26, 33, 34, 112, 125, 129, 131, 134, 143, 161, 173, 184, 188, 225
Dennett, D., 150
Derefeldt, G., 280
DeValois, K. K., 92, 107
DeValois, R. L., 92, 107
Diamond, A. L., 52
du Buf, J. M. H., 52, 63, 70, 90
D'Zmura, M., 202

E

Eisner, A., 81
Engvik, H., 91
Enroth-Cugell, C., 98, 162, 182, 183, 184, 212
Epstein, W., 221, 258, 273
Evans, R. M., 37, 41, 93, 96, 129, 132, 133, 143, 144, 145, 161, 173

F

Fach, C., 49
Fairchild, M. D., 211
Festinger, L., 137
Fiorentini, A., 36
Flock, H. R., 8, 21
Forsyth, D. A., 161, 169, 200
Freeman, R. B., 14
Friden, T. P., 20
Fry, G. A., 194
Frykholm, G., 265
Fuchs, W., 280

G

Gelb, A., 10, 19, 265
Georgeson, M. A., 52
Gerbino, W., 11, 28, 199, 255, 285
Gerrits, H. J. M., 105
Gershon, G., 202
Gibson, J. J., 32, 215, 216, 217
Gilchrist, A. L., 2, 5, 6, 8, 11, 15, 16, 17, 19, 21, 22, 23, 26, 32, 33, 34, 39, 60, 112, 115, 117, 118, 125, 129, 131, 133, 134, 136, 137, 141, 142, 143, 144, 159, 161, 173, 181, 184, 188, 223, 240, 254, 273, 274, 275, 276, 279, 280, 281, 286
Glad, A., 69, 91, 92

Godlove, I. H., 65
Gogel, W. C., 277, 280
Goldstein, R., 6, 11, 16, 33, 106, 113, 115, 121, 125, 134, 160, 163, 164, 166, 167, 179, 191, 192, 193, 194, 198, 209, 210, 272
Gottesman, J., 63, 70
Graham, M. E., 100
Graham, N. V. S., 157
Grindley, G. C., 132
Grossberg, S., 194
Gustafsson, K. A., 173, 257, 258, 266, 267, 269, 270, 277, 278, 280, 285, 286

H

Haimson, B. R., 21
Hallett, P., 202
Hamada, J., 37
Hasegawa, T., 212
Hatfield, G., 221
Hayhoe, M., 56, 98, 102, 104, 105, 107, 212
Heeger, D. J., 86
Heggelund, P., 81, 90, 145
Heider, G. M., 228
Heinemann, E. G., 7, 14, 72, 81, 116, 117, 125
Helmholtz, H., 18, 37, 215, 228, 229
Helson, H., 18, 20, 93, 119, 120, 125, 136, 137, 142, 201, 202, 259
Henneman, R. H., 16, 111, 115
Hering, E., 19, 20, 41, 259
Herschberger, W., 278
Hess, C., 26, 36, 78, 117
Hess, E. H., 278
Hochberg, J. E., 273
Holliday, I., 97, 138, 139
Holmgren, S., 273
Hood, D. C., 98, 102, 104, 105, 107
Horn, B. K. P., 33, 161, 169, 170, 173, 179, 194
Howard, I., 278
Howell, E. R., 92, 138
Hsia, Y., 111, 266
Hunt, R. W. G., 211
Hurvich, L. M., 8, 14, 15, 17, 18, 20, 21

I

Ivry, R., 217, 218, 227, 231, 239, 243, 245, 246, 247, 250

J

Jacobsen, A., 2, 6, 8, 11, 16, 19, 21, 23, 26, 32, 33, 34, 60, 112, 115, 117, 118, 125, 129, 131, 134, 142, 143, 161, 173, 181, 184, 188, 225, 240, 254

Jakobsson, T., 258, 277, 285
Jameson, D., 8, 14, 15, 17, 18, 20, 21
Jeffers, V. B., 259
Jepson, A., 202
Johansson, G., 29, 173, 220, 259, 260, 265
Johnstone, J. R., 192
Judd, D. B., 93, 162, 182, 201, 202, 259
Jung, R., 38, 89

K

Kanisza, G., 219, 221
Kardos, L., 266
Katz, D., 10, 172, 257, 265, 280, 286
Keemink, C. J., 192
Kelly, D. H., 5
Kersten, D., 63, 70
Kingdom, F., 37, 88
Kinney, J. A. S., 212
Klinker, G. J., 200, 211
Koenderink, J. J., 86
Koffka, K., 3, 10, 16, 17, 19, 32, 172, 188, 216, 219, 228
Köhler, W., 10, 19
Koshel, R. J., 105
Kozaki, A., 173
Kozaki, T., 52
Krauskopf, J., 26, 30, 116
Kronauer, R. E., 94

L

Laming, D. R. J., 72, 86, 134
Land, E. H., 25, 33, 95, 179, 184, 191, 259, 262, 263, 266
Lange-Malecki, B., 202
Lasley, D. J., 76
Le Grand, Y., 68, 72
Leeuwenberg, E. L. J., 221
Legge, G. E., 63, 70
Lennie, P., 202, 211
Leshowitz, B., 76
Levin, M. E., 105
Lie, I., 113, 173
Lingelbach, B., 92, 107
Lockhead, G. R., 25, 136, 138, 192
Lucassen, M. P., 94
Luccio, R., 221
Lynch, E., 212

M

MacAdam, D. L., 210
Mach, E., 35, 259, 273
MacLeod, D. I. A., 45, 53, 56, 57, 81, 98
MacLeod, R. B., 18, 22, 112

Magnussen, S., 36, 52, 62, 69, 91, 92
Mallot, H. A., 199
Maloney, L. T., 21, 33, 179, 204, 205, 206, 207, 211
Marks, L. E., 81
Marmolin, H., 265
Marr, D., 33, 37, 150, 169, 170, 177
Masin, S. C., 224
Masuda, N., 21
Mausfeld, R., 21, 41, 93
Maximov, V. V., 200
McCann, J. J., 25, 33, 95, 163, 179, 184, 191, 201, 202, 259, 262, 263, 266
McKee, S. P., 201, 202
Merleau-Ponty, M., 149
Metelli, F., 217, 218, 219, 227, 228, 230, 238, 239, 257, 271, 280, 286
Metzger, W., 259
Michels, W. C., 259
Mingolla, E., 198
Moulden, B., 37, 88
Moustgaard, I. K., 278
Mueller, C. G., 38
Mullen, K. T., 97
Munsell, A. E. O., 65

N

Nachmias, J., 72
Nakayama, K., 144
Nayatani, Y., 208
Neisser, U., 12
Neri, D., 212
Newhall, S. M., 133
Niederee, R., 21, 41, 93
Nikolaev, P. P., 5, 200
Noguchi, K., 21
Nordby, K., 49, 53
Nuboer, W., 21, 129

O

O'Connell, D. N., 259
Ogawa, T., 52
Ohmi, M., 278
Okayama, K., 52
Oppel, J. J., 278

P

Parrish, M., 132, 144
Pentland, A. P., 29, 143, 144
Piantanida, T., 26, 141
Pokorny, J., 201
Poulton, E. C., 65

Prazdny, K., 217, 218, 227, 231, 239, 243, 245, 246, 247, 250
Pretori, H., 7, 8, 26, 36, 117
Putaansuu, J., 173, 192, 257, 266, 267, 269, 270, 277, 278

R

Raab, D. H., 76
Ramachandran, V. S., 144
Reeves, A., 167, 202, 209, 211
Reid, R. C., 136, 137, 138, 140
Remondino, C., 232
Restle, F., 221
Rittenhouse, D., 278
Rivers, G., 137
Rivers, W. H. R., 142
Robson, J. G., 101
Rock, I., 10, 215, 220, 273
Rogers, B. J., 100
Rogowitz, B., 94
Ronchi, V., 216
Ross, J., 192
Rozhkova, G. I., 5
Runesson, S., 265

S

Sällström, P., 204
Schechtman, G., 86
Schiller, P. H., 36
Schirillo, J., 167, 209
Schlosberg, H., 258
Schnapf, J., 98
Schröder, H., 278
Schubert, J., 56
Semmelroth, C. C., 65, 75, 189, 190
Sewall, L., 153
Shafer, S. A., 200
Shapley, R., 98, 136, 137, 138, 140, 162, 182, 183, 212
Sharkey, T. J., 281
Sharpe, L. T., 49, 53
Shchardrin, V. E., 5
Shevell, S. K., 84, 94, 97, 106, 138, 139, 201
Shimojo, S., 144
Sloan, L. L., 65
Smith, K., 133, 144
Smith, V. C., 201
Sobagaki, H., 208
Solomon, J. A., 100
Spehar, B., 115, 121, 122, 125, 130, 160, 185, 186
Spencer, D. E., 93
Sperling, G., 100
Spillmann, L., 65

Steinman, R. M., 72
Stevens, K. A., 37
Stevens, S., 8
Stiles, W. S., 44, 46, 48, 65, 96, 116, 161, 182
Stockman, A., 45, 49, 53, 56, 57
Stromeyer, C. F. I., 94
Stultiens, C., 11, 199, 255, 285
Sullivan, G. D., 52
Swanston, M. T., 71, 72, 79

T

Takahama, K., 208
Takano, Y., 52
Takasaki, H., 65, 188
Taub, H. B., 76
Taylor, T. H., 201, 202
Teller, D. Y., 72
Tenenbaum, J. M., 12, 13, 106, 143, 168, 169, 170
Thijssen, J. M., 79
Thomas, J. P., 36
Tietz, J. D., 281
Timberlake, G., 5
Todd, J. T., 31, 198
Todorovic, D., 198
Troost, J., 11, 189, 255, 285
Tuijl, H. F. J. M., 221

U

Ullman, S., 86

V

Valberg, A., 202, 212
van de Grind, W. A., 86
van den Brink, G., 192
van der Helm, P. A., 221
van der Wildt, G. J., 192
Vendrik, A. J. H., 79, 105
Verillo, R. T., 5
von Fieandt, K., 278
von Hofsten, C., 265
von Kries, J., 45
Vos, J. J., 94

W

Wagman, I. W., 105
Wallach, H., 4, 7, 26, 36, 222, 254, 259
Walls, G. L., 93, 105
Walraven, J., 12, 29, 30, 43, 93, 94, 98, 142
Wandell, B. A., 21, 33, 129, 202
Ware, C., 212
Warren, M. H., 201

Waygood, M., 5
Webster, M. A., 92, 107
Wenderoth, P., 56, 104, 212
Werner, J. S., 65
Wertheimer, M., 13, 24
Whiteside, J. A., 194
Whittle, P., 6, 21, 22, 25, 26, 29, 37, 38, 39, 43, 45, 46, 48, 53, 56, 57, 62, 65, 66, 70, 71, 72, 74, 75, 76, 79, 80, 81, 83, 87, 88, 90, 92, 93, 94, 95, 97, 99, 104, 105, 116, 126, 132, 133, 136, 138, 139, 141, 143, 145, 159, 173, 182, 188, 190, 191
Williams, L. G., 88
Wilson, M. H., 93
Wittgenstein, L., 253
Woodworth, R. S., 37, 128, 129, 142, 143, 144, 258
Wooten, B. R., 143
Worthey, J. A., 161, 201, 202
Wyszecki, G., 44, 46, 48, 65, 96, 116, 161, 162, 182, 200, 201, 202

Y

Yamanaka, T., 208
Yarbus, A. L., 5, 25, 26, 141
Young, T., 94, 107

Z

Zisserman, A., 161, 169, 200
Zucker, S. W., 168

Subject Index

A

Absolute luminance, 3–8, 16, 18, 21, 27, 136, 184
Achromatic color, 203–205
Adaptation, *see also* Gain, Self-adaptation, 3, 18, 161–162, 179, 182–188, 211–212
  light, 43, 102–105
  rapid, 98
Adaptation-level theory, 10, 18
Assimilation, 136–141
Attenuation, *see* Gain
Attitude of observer, 10, 19, 126, 132, 218

B

Black limit, 58, 74
Brightness, *see also* Contrast brightness, Matching, 5–7, 10, 11, 16, 18, 25, 32, 35, 162, 183–188
  area, 52
  of colored lights, 48, 81
  constancy, *see* Lightness constancy
  duration, 50, 62
  functions, 65, 81
  versus lightness, 10, 18, 113
  matching, *see* Matching, of brightness
  mesopic, 50, 118–119
  physical contrast, 68, 111
  in rod vision, 48
  S-cones, 81
Brightness matching, *see* Matching

C

Center-surround
  receptive field, 153
  stimulus pattern, 4–6, 11, 14, 15, 25–27, 29–31, 33, 39, 184–186, 188–190, 195, 198
Color
  chromatic, 29, 199–212, *see also* Contrast color
  constancy, *see* Constancy, of color
  luminous, *see also* Modes of appearance, 283
  rendering, 282, 286
Common and relative components, 4, 9, 29–31, 34, 100, 223–224, 231, 237, 253, 257, 260, 262, 265–266, 280, 282–284, 286
Commonality assumption, 257, 262, 265, 280, 286
Complexity of image, 29, 33, 160–166, 179, 191, 211, 223, 225
Constancy
  of color, 10, 32, 199–212, 257, 259, 262, 265–266, 272, 278, 280–282, 286
  failure of, 20, 69
  hypothesis, 3, 17
  of lightness, *see also* Lightness constancy, 240, 253, 255, 272, 286
  mode, *see also* Attitude of observer, 10
  in projected slides, 127
Contour, *see* Edge
Contrast, *see also* Contrast theory, Physical contrast, Simultaneous contrast,

315

14–17, 35–36, 159–162, 179, 181, 182, 185–194, 196, 201, 212–213
apparent, 63
cone, 94
Contrast brightness, *see also* Luminance discrimination, 35–112, 141–151
  vs. apparent contrast, 108
  conditions for, 141–145
  contour maps of, 67–71, 81–84
  converging evidence, 77, 109
  cross-study comparisons, 115–125
  dynamic, 36, 91, 284
  intransitivity, 147
  phenomenology of, 37, 145
  in rod vision, 48
  and stereo depth, 37, 100
  two-stage model, 107
Contrast-coding, 2–5, 7–8, 22, 38, 84–86, 98, 135, 159–162, 181–188
  generality of, 86, 99
  retina vs. brain, 106
Contrast color, 92, 92–97, 151–152
Cornsweet effect, *see* Craik–O'Brien effect
Craik–O'Brien effect, 25, 138, 192–193, 284–285
Crispening effect, 65, 74, 135

D

Darkness, *see also* Decrements, 57
Decoding Principles, *see* Perceptual organization
Decrements, 35, 57–62, 186
  discrimination, 73–77
  vs. increments, 6–8, 30, 36, 57, 88, 186
  of surround, 64
Depth, *see also* Contrast brightness, 29, 34, 166, 177, 192, 199, 216, 272–273, 276–281
Differencing, 98
  spatial vs. temporal, 101
Discounting, 191
  background, 21–22, 43, 93, 100, 147
  and differencing, 100, 102
  illumination, 21, 24, 32–33, 93
Discrimination experiments, *see* Luminance, discrimination
Disk-annulus, *see* Center-surround

E

Edge, *see also* Illumination border, Reflectance border, 2–5, 14, 24, 29–31, 38, 101, 163, 173, 180–181, 184, 191–192, 210, 239–240

classification, 2, 4, 9–10, 16–17, 23, 31–34, 134, 143, 173, 184, 210
integration, 2, 4, 6–8, 12, 16, 24–27, 134, 139, 163, 173, 180, 210
maintaining a gestalt, 147, 258, 284, 286
Encoding, *see* Contrast-coding
Energy vs. information, 3, 8, 10, 13

F

Fechner's integration, 71, 78
Figure-ground organization, 147
Filling in, 105
Fourier analysis, 12–14

G

Gain
  exponential, 45
  multiplicative, 45, 98
Ganzfeld, 5, 8
Gelb effect, 9–10, 12, 266
Gradient, 16, 159, 161–166, 173, 176–183, 191–194, 200–202, 209–213
  analysis of, *see* Vector analysis
  of apparent illumination, 163
  of illumination, 17, 19, 26, 32–33, 163, 200–202, 209–213, 260, 277, 284
Gray, 86
Grating, 266–268, 277–279

H

Helson–Judd effect, 96
Holes, *see also* Black limit, Decrements
  perception of darkness, 131
HSD, 41, 90, 139, 141

I

Ilco, 257, 262–263, 283
Illumination, 161–166, 173, 176–183, 191–194, 200–202, 209–213, 221, 223–228, 231, 238, 240
  perceived direction of, 277–278
  perception of 2, 8–13, 18–21, 24, 28–29, 32–33, 127, 131, 262, 284–285
Illumination border, 19, 23, 26, 32–34, 161–166, 173, 176–183, 191–194, 200–202, 209–213, 239–240, 257, 259, 262, 284–286
Increments, *see also* Decrements, vs. increments, 35, 43–57, 186
  of surround, 64
Insistence, 10

SUBJECT INDEX   317

Integration, *see also* Edge, integration, 126, 134–136
Intrinsic images, 8, 12, 19, 29, 160, 166–213
Introspectionism, 9
Iso-brightness curves
  decrements, 58–64
  and discriminability, *see* Fechner's integration
  equations of, 46, 49, 61
  increments, 41–57
  scotopic, 49

L

Lateral inhibition, 3–4, 9, 15–16, 18, 23, 153
Lighting, *see* Illumination
Lightness, 10, 35, 162
  vs. brightness, 113, 133, 161–162
  constancy, 3, 9, 11, 15–26, 32–34, 69, 115, 119–125, 159–198
  two kinds of, 21–22, 126–133
  with respect to illumination, 126–131
  with respect to surround, 24–25, 126–131
Local vs. global determination, 2–4, 24, 33, 143
Luminance
  and contrast brightness, 73–79
  discrimination, 71–77
  profile, 262, 264, 267, 282
  ratio, *see also* Contrast, Ratio principle, 5–7, 11, 15–16, 18, 22, 24–25, 259, 271, 275, 285

M

Matching
  brightness, 7, 11, 16, 26–27, 39–64, 183–188
  color, 95
  contrast, 11, 63, 183–188
  illumination, 11, 127
  lightness, 11, 16, 20, 22, 26–27, 183–188, 265, 273, 275–276, 279
  luminances, 6–7, 183–188
  ratios, 6–7, 11, 183–188
  transparency, 11
Mesopic vision, *see* Brightness
Michelson contrast, 70
Minimum principle, 28–29, 257, 272, 285–286
Modes of appearance, 10–11, 146, 283
Modulation depth, *see also* Physical contrast, 266
Mondrian, 4, 11, 33, 121, 163–166, 185–188, 198–200, 202, 210–2
Mosaic conception, 3, 12–13, 30
Multidimensional appearance, 2–3, 8–13, 24, 28, 31–32, 166–213

O

O'Brien effect, *see* Craik–O'Brien effect
Organization, *see* Perceptual organization

P

Perceptual organization, *see also* Figure-ground, 2, 17, 19, 23, 28, 33–34, 281–282
Perceived layers, 3, 9, 12–13, 17, 28–31, 166–213
Percept–percept relations, 280–282
Phenomenal experience, 7, 9, 13, 20, 23, 194–197
Photometer metaphor, 1, 3, 5–8, 12–13, 17–24, 32, 39
Physical contrast, *see also* Michelson contrast, 4–5, 7, 11, 13–15
  modulation depth, 91, 270, 278–279
Pi-mechanisms, 46–48, 94
Point-by-point representation, 3–4, 9, 13
Pronouncedness, 10
Proximal mode, 10

Q

Qualia, *see* Sensations

R

Ratio matching, *see* Matching
Ratio principle, 5–7, 15, 18, 21–24, 29, 33, 36
Reco, 257, 262–263
Reflectance border, 19, 23, 32–34, 159–198, 216–218, 239–240, 246, 257, 262, 284–286
Relational approach, 2–4, 12
Relative component, *see* Common and relative components
Relative luminance, 13–16, 27, 30
Retinex theory, 95, 134, 163–165, 259

S

Scaling, *see* Contrast brightness
Scattered light, 53, 62
Scotopic vision, *see* Brightness
Self-adaptation, 56
Sensations, 150–152
Shadow, 127, 257, 262, 270–272
  attached, 19, 266, 272, 277, 281
  cast, 9, 12, 34, 272
Shape from shading, *see* Solid shape
Simultaneous contrast, *see also* Contrast brightness, 35, 188–191

318   SUBJECT INDEX

brightness, 14, 22, 36, 38, 69, 116–119, 132, 152, 188–191
color, 212
edges, 38, 101
lightness, 14, 22, 25–26, 153, 188–191
lightness constancy, 15, 69, 115, 188–191
Solid shape, 19, 166–180, 192–193, 199–200, 272
Spatial frequency analysis, *see* Fourier analysis
Specfic component, *see* Common and relative components
Stabilized retinal image, 5, 21, 24–26, 31, 141
Stiles–Crawford effect, 48
Stimulus pattern, *see also* Center-surround, Grating, HSD, Mondrian, 257, 262–263, 266, 268–270
Structure, *see* Perceptual organization
Suprathreshold psychophysics, 48
Surround
    vs. background, 93
    effect on brightness, 83
    effect on lightness, *see also* Lightness constancy, 266
    flashing, 64

T

Temporal modulation, *see* Contrast brightness, dynamic
Threshold, 33, 43
    t.v.i. function, 44
    two-color, 46
Transparency, 11–12, 16, 28–29, 215–255

V

Vector analysis, *see also* Perceptual organization, 161–166, 173, 176–183, 191–194, 200–202, 209–213, 220
    of contrast color, 93
    of reflected light, 29, 34, 257, 283, 285
Veiling luminance, 9, 11, 17, 21, 31, 240
Venn diagram, 9
Von Kries adaptation, 44, 161

W

W (L/L_{min}), 41, 58, 65, 69, 76, 87

LIBRARY   OPTICS/LIGHT/COLOR
copy #1   ACC # 8606

GILCHRIST (60.)

Exploratorium Library
3601 Lyon Street
San Francisco, CA 94123
DISCARDED
Exploratorium
Learning Studio

GAYLORD